The Mathematical Sciences

A Collection of Essays

Published for the
National Academy of Sciences—National Research Council by
The M.I.T. Press
MASSACHUSETTS INSTITUTE OF TECHNOLOGY
CAMBRIDGE, MASSACHUSETTS, AND LONDON, ENGLAND

The Mathematical Sciences

A Collection of Essays

Edited by the National Research Council's
Committee on Support of Research in the Mathematical Sciences (COSRIMS)
with the collaboration of George A. W. Boehm

Set in Monotype Baskerville and printed and bound in the United States of America by The Colonial Press Inc.

Designed by Dwight E. Agner

Library of Congress catalog card number: 69–12750

Foreword

The Committee on Support of Research in the Mathematical Sciences (COSRIMS) was appointed by the Division of Mathematical Sciences of the National Research Council at the instigation of the Committee on Science and Public Policy of the National Academy of Sciences.

The composition of COSRIMS was as follows:

Lipman Bers, *Columbia University*, Chairman
T. W. Anderson, *Columbia University*
R. H. Bing, *University of Wisconsin*
Hendrik W. Bode, *Bell Telephone Laboratories*
R. P. Dilworth, *California Institute of Technology*
George E. Forsythe, *Stanford University*
Mark Kac, *Rockefeller University*
C. C. Lin, *Massachusetts Institute of Technology*
John W. Tukey, *Princeton University*
F. J. Weyl, *National Academy of Sciences*
Hassler Whitney, *Institute for Advanced Study*
C. N. Yang, *State University of New York at Stony Brook*
Truman Botts, *University of Virginia*, Executive Director

Our task was to assess the present status and the projected future needs, especially fiscal needs, of the mathematical sciences. It was clear to us from the very beginning of our work that our report would have to differ somewhat in structure from the corresponding reports for other disciplines that had already appeared.

Though mathematics provides the common language for all sciences, we realize that even scientific readers of our report, let alone nonscientists, may feel that they are not adequately informed about what mathematical research, especially modern mathematical research, consists of. Similarly, even professional mathematicians, or scientists who customarily use mathematics in their work, may be unaware of the manifold applications of mathematics in various sciences and technologies, especially the new applications influenced by the computer revolution.

To provide additional background of factual information concerning the mathematical sciences, we are supplementing our report[1] with the present collection of essays, written by distinguished authors on various topics in mathematics, in the applied mathematical sciences, and in the applications of mathematics. With three exceptions, which are reprints, these essays were written expressly for this collection.[2] They are intended not only for the nonmathematical scientist but also for the scientifically oriented layman. Opinions expressed in an essay are of course those of its author and not necessarily those of the Committee or the National Research Council.

Mathematics pervades our whole educational system. As a matter of fact, we believe that the mathematical community has no obligations more important than those concerned with education, the most critical area being collegiate education. We have, therefore, included in our report questions of policy regarding higher education. Our Panel on Undergraduate Education has carried out an intensive study of this area; its findings are presented in more detail in a separate volume.[3]

Simultaneously with our activities, the Conference Board of the Mathematical Sciences has been carrying out a survey of research and education in mathematics, and its survey committee has agreed to act as a fact-finding agency for our committee. The Conference Board Survey Committee's report[4] will contain a wealth of factual and statistical material pertaining to the matters discussed in our report. We take this opportunity to express our gratitude to the Survey Committee and to the Ford Foundation, which supported their work.

[1] *The Mathematical Sciences: A Report*, by the Committee on Support of Research in the Mathematical Sciences, National Academy of Sciences (1968).

[2] The three exceptions are Dyson's essay, which first appeared in the *Scientific American*, Vol. *211*, No. 3 (1964), p. 129; Gleason's essay, which first appeared under the title "Evolution of an Active Mathematical Theory," in *Science*, Vol. *145* (1964), p. 451; and Kac's essay, which is a modification and expansion of his essay in the *Scientific American*, Vol. *211*, No. 3 (1964), p. 92. We are grateful to both the *Scientific American* and *Science* for permission to include these three essays in the form in which they appear here.

[3] *The Mathematical Sciences: Undergraduate Education*, by the Panel on Undergraduate Education of the Committee on Support of Research in the Mathematical Sciences, National Academy of Sciences (1968).

[4] Volume I of this report, *Aspects of Undergraduate Training in the Mathematical Sciences*, Conference Board of the Mathematical Sciences (1967), has already appeared.

The activities of our committee have been financed mainly by a grant from the National Science Foundation. This has been supplemented by smaller grants from the Sloan Foundation, the Conference Board of the Mathematical Sciences, the American Mathematical Society, the Association for Computing Machinery, the Association for Symbolic Logic, the Institute of Mathematical Statistics, the Mathematical Association of America, the National Council of Teachers of Mathematics, the Operations Research Society of America, and the Society for Industrial and Applied Mathematics. Columbia University has generously provided us with office space and many auxiliary services. To all these organizations we express our thanks.

We are deeply indebted to the authors of the essays, to the chairmen and members of our panels, and to the many other individuals who have contributed their time and expertise to our undertaking.

January 20, 1968

Lipman Bers
Chairman, Committee on Support
of Research in the Mathematical Sciences

Contents

The Mathematical Sciences

A Collection of Essays

How would you define "useful" mathematics and what would you do to get more of it invented? This question, in one form or another, worries many research directors and other senior statesmen in engineering and sciences. This essay, in scanning the varied uses of mathematics today, suggests that there is no direct path between the invention of new mathematics and its applications. Throughout the long history of mathematics, ideas that seemed to be little more than frivolous fantasies when first discussed have proved ultimately to be applicable to a variety of mundane but important problems. Although much of the mathematics born today will no doubt prove similarly serviceable, its probable uses cannot be easily identified at an early stage of development.

The Applicability of Mathematics

Stanislaw Ulam

Current research in mathematics tends toward ever-more-varied abstraction. Yet the most far-reaching excursions into mathematical theory may lead to applications not only within mathematics itself but also in physics and the natural sciences in general. While it is true that much of the work published in mathematical journals consists of detailed and rather specialized investigations, one might think of them as "patrols" sent into the unknown in all directions. Some of these turn out to encounter new areas of interest in the great game that the human mind plays with nature. Even though much of the activity in mathematical research may appear to be "getting off on tangents," the great sphere of knowledge is increasing at an important rate, and the applicability of mathematics to problems suggested by new discoveries seems to know no boundaries.

Mathematics as we now know it has historically a twofold origin. In antiquity the beginnings of two modes of mathematical thought arose almost simultaneously and probably independently: the arithmetic one — the wondering about numbers — and the geometric one, concerned with figures. It is perhaps remarkable that on the whole the same persons cultivated both these modes of thought. The idea of examining the thought process itself, that is, becoming conscious of logic and of the mathematical method, probably came somewhat later. A certain dichotomy persists in mathematical work even at present. The mathematical *method*, as presently used, probably would not appear strange to the Greeks. However, the *objects* to which mathematical thought is devoted today have been vastly diversified and generalized. It is their proliferation that would perhaps appear so striking not only to the ancients but even to mathematicians of the last century.

The manifest usefulness of applying the notion of numbers to geometric objects, for example, in ideas of length, area, volume, can serve as a first example of the applicability of mathematics. If we consider the beginnings of geometry as observational work concerning the properties of physical reality rather than as exercises in pure thought, one might say that already in the prehistory of mathematics it is hard to make precise distinctions, epistemologically, between pure and applied mathematics. And yet the different psychological motivation and the different emphasis in method allow us to talk meaningfully of this distinction. Perhaps mathematics is unique among activities of the human mind in being, on the one hand, so much an art for art's sake and providing, on the other, so many tangible applications that change the course of the human condition. But in turn these applications mold to a certain extent the development of this abstract art itself.

Philosophically it is equally curious that mathematical idealizations which at first sight seemed "irrational" have led to the most useful and practical consequences. It is the idea of infinity both "in the large" and in the "infinitesimally small" that is so overwhelmingly successful. *A priori*, it is not obvious why the algorithms of the calculus — compared with the operations with finite differences that are more palpably suggested by the first experiences with numerical quantities — should be so convenient, elegant, and powerful in their use. The many applications of the infinitesimal calculus in the natural sciences, especially in astronomy and physics, are to this day perhaps the greatest triumph of mathematical thought. As for the "infinities in the large," the laws of large numbers in probability theory and statistical mechanics allow for more convenient and penetrating formulations than any systems of finite inequalities. Even in pure mathematics itself (for example, in number theory), the asymptotic theorems reveal regularities and give better insight than the more "local" theorems. The ergodic theorems give equally good insight into the properties of many mathematical and physical phenomena.

It is impossible to conceive of present-day technology or, indeed, the development of the exact sciences and the technology of the nineteenth century without the previous invention and availability of the infinitesimal calculus, beginning with its role in celestial mechanics and astronomy and followed by its penetration of all of classical physics and its involvement in most practical engineering achievements. The achievements of electricity and magnetism are equally unthinkable without the expanded apparatus of mathematical analysis which permits continuous fields to be described in an adequate fashion. At the same time, some of the mathematical ideas of the nineteenth century, such as those of the non-Euclidean geometries, found their role and application in physical sciences only in the present century. One knows the importance of the ideas of Riemann in the theory of relativity. The second half of the nineteenth century was especially fertile in purely mathematical constructions that have found and are increasingly finding applications in physics.

The Use of New Ideas

In this little article, we shall try to select, somewhat arbitrarily, illustrations of some of these applications, and we shall attempt to give examples of the similar use of mathematical ideas of the twentieth century up to the present time. The great quantum step in the development of the mathematical outlook made by Cantor's set theory has not only affected profoundly the foundations of mathematics but has enlarged enormously the generality of mathematical objects. The theory of very abstract sets leads in natural fashion to consideration of more general structures to serve as spaces for geometrical studies. It was equally natural to consider more general functions than the ones forming the primary objects of classical analysis, for example, arbitrary continuous functions without the conditions of differentiability or analyticity previously imposed on them. The preponderant study of functions of real or complex variables was enlarged to form a study of transformations of spaces. It was equally natural to consider classes of functions as themselves forming spaces, endowed with geometries of their own. Functional analysis became an impressive edifice of mathematical work.

The generality of the idea of space and the successful employment of these notions stimulated a parallel development of very general abstract algebraical structures. All this very extensive work can be regarded as an application of new mathematical abstractions to the mathematical objects previously considered. One striking example is the use of fixed-point theorems, generalized from finite dimensional Euclidean spaces to spaces of functions. These are important in establishing the existence of solutions of differential or integral equations. As another example, we may mention ergodic theorems. These theorems, proved in a framework of the theory of measure of sets and of real variable theory, lead to theorems like the laws of large numbers in the theory of probability. One could give numerous other examples of the applicability of the more general mathematical methods to older, more "classical" problems. The more abstract and purely combinatorial thinking was found to be of great use in problems of classical number theory. This is evidenced, for example, by the work of Schnirelman on the problem of representing all integers as sums of a fixed number of primes.

The theory of groups developed from the study of purely algebraic problems by Abel, Galois, and their successors. In the hands of F. Klein, this development found a programmatic role in the foundations and planning of geometric theories. But in the course of the last few decades the theory of groups has begun to play an increasingly important role in the very foundations of physical theories. It has provided most important tools for the ordering and the classification of atomic spectra. It is through group theory that the apparent chaos of the spectral lines can be replaced by an order deriving from general principles of quantum theory. More fundamentally important still, starting at the beginning of this century the special theory of relativity was able to impose its all-embracing role through use of the notion of the group of transformations under which physical laws have to be invariant. The important thing to remember is that not only the transformations of the Minkowski spaces in themselves but the abstract properties of the group which they constitute were essential for this purpose. The Lorentz group is thus one of the most important ideas in all of mathematical physics.

In more recent times, the notions of group theory have become very useful and perhaps crucial for the understanding of the variety of "fundamental" particles. We cannot describe in our limited space the origin and the scope of these ideas but shall have to content ourselves with mentioning that at the present time important work is proceeding, to account for the variety and properties of the elementary particles through assumptions of the existence of a few groups of symmetries governing, as it were, the choices of nature. In addition to the Lorentz group, these symmetries seem to involve certain more finitely describable ones, for example, symmetries of the spin, the charge, and the duality between particles and antiparticles.

It is quite well known that the development and the use of mathematical tools were a necessary prerequisite, and indeed often a stimulus, to the development of present-day technology. The language and the technique of calculus permitted not only the developments of the machine age, such as the construction of bridges and the design of electric motors, but also the formulation of theories of thermodynamics, electric fields, chemical reactions, flight in the atmosphere, and so on. The recent achievements in rocket propulsion and the construction of artificial satellites are all based on principles of classical mechanics, a vast field of application of mathematical analysis to consequences of Newton's laws. So much of all this is taken for granted that we find it worth while to remind ourselves of this obvious and common role of mathematics.

Perhaps not so well known is the role of mathematical thinking and mathematical techniques in some other technological achievements of this century. It should be quite obvious that the atomic era could come only after the great discoveries of theoretical physics — of the equivalence of mass and energy, which followed from the theory of relativity — and the better understanding of properties of matter in general, which came through quantum theory. For the construction and technology of the atomic reactors, the apparatus of mathematical physics was essential. In the construction of atomic bombs and of the hydrogen bomb, an enormous amount of mathematical work had to be done. This involved not only the techniques of classical analysis but also some of the more recent mathematical developments that made possible modern thermodynamics: statistical mechanics of material particles and of fields of radiation.

The theory of probability pursued by mathematicians in this century and the theory of spaces with infinitely many dimensions, like Hilbert space, have found applications. In the work now proceeding on the construction of fusion reactors (that is, the attempt to design devices which, in contrast to the H-bomb, would release the energy gradually), an enormous amount of mathematics of the most sophisticated type is being used. It is very interesting to some "pure" mathematicians to see how some very abstract-looking theorems (for example, on the existence of fixed points of transformations and of periodic orbits) find important applications in the design of the big accelerating machines propelling protons or electrons to velocities very close to that of light. The increasing complexity of not only the problems posed by technology but the formulations of the very foundations of physics makes an increasing demand on the use of the most modern and complex mathematical ideas. The problems to which the most advanced mathematical ideas are being applied include such ambitious enterprises as that of calculating the general circulation of the atmosphere of our globe and of making weather predictions. Mathematically speaking, this involves systems of partial differential equations in three spatial dimensions and in the time variable. Mathematical work has been greatly expanded and is now proceeding on the understanding of the qualitative and quantitative behavior of the solutions of such systems.

New Paths to Biology

In very recent times another area of the natural sciences appears to be becoming mathematizable. This is the great field of molecular biology. The recent fundamental discoveries in this field begin to form a framework for theories promising an understanding of the life processes. The fundamental processes of replication of a living cell, the code describing the construction of materials necessary for what we call life and its fundamental processes, begin to emerge as a rationally describable set of schemata. It is hoped that the ideas of structures of mathematical systems in general and especially the field of mathematics known as combinatorial analysis will be useful in working out the perplexing complexity of these arrangements.

The future understanding of the function of the nervous system in living organisms and some of the mysteries involved in the workings of the human brain itself may be gradually aided by new mathematical ideas and techniques.

In all these new problems, presently under full study, a new tool has been and will increasingly be of decisive use: electronic computing machines. As we mentioned before, the new and more ambitious problems to which mathematics is applied are characterized by a complexity far exceeding that encountered in previous problems of physics and technology. The speed with which computers operate allows performance of the billions of operations, necessary to computation of many such complex phenomena. The very recent achievements in space technology, like the rendezvous between space vehicles, are unthinkable without the availability of both the telemetry and the electronic computing machines, which together permit almost instantaneous calculation of the changes in propulsion necessary for the achievement of prescribed orbits. In the calculation of the motion of air masses for weather forecastings, the computers are not only useful but absolutely necessary. Beyond all this, the designing and the operation of the electronic computers themselves involve ideas of mathematical logic and combinatorial analysis, developments of the present century and of the present time. Studies of these methods, too, promise to be of use in the problems of the functioning of the nervous system and of the human brain.

BIOGRAPHICAL NOTE

Stanislaw Ulam is both a research advisor at the University of California's Los Alamos Scientific Laboratory in New Mexico and a professor at the University of Colorado. Born and educated in Poland, he came to the United States in 1936. During World War II he was involved in work on the Manhattan Project at Los Alamos, where he later became one of the intellectual parents of thermonuclear weapons. He is also well known for pioneer work on the Monte Carlo Method, a widely used procedure for solving problems in mathematics and physics by simulating them repeatedly with computers. Ulam's interest in applications of mathematics extends to many diverse fields, including biology and advanced methods of nuclear propulsion for space travel.

Complex analysis arose from a dilemma that confronted Italian mathematicians during the Renaissance. For coping with certain problems in algebra, they needed to make use of the square root of -1. Yet it went against the grain to admit into the family of numbers this "imaginary" unit, which has no value that can be expressed by ordinary real numbers. Gradually mathematicians overcame their prejudice against imaginary numbers and built up a distinct body of mathematics around them. They learned that the square root of -1 could be combined with real numbers and that the result, complex numbers, obeyed the laws of ordinary algebra and arithmetic. Today complex analysis is one of the universal tools in both pure mathematics and applications to science and engineering. It is the key to research on prime numbers. And it is applied to diverse problems in aeronautics, liquid flow, the design of electronic circuits, and many other fields. This essay sketches the development of complex analysis and some of its uses, past and present.

Complex Analysis

Lipman Bers

One of the first purely mathematical problems ever considered was the solution of quadratic equations. The technique of solving such equations was discovered in ancient times in Babylonia; it is essentially the same technique as is taught today in high schools. To find a number x such that $x^2 - 2x - 15 = 0$, we write the equation in the form $x^2 - 2x + 1 - 16$, which is the same as $(x - 1)^2 - 16 = 0$ and conclude that $(x - 1)^2 = 16$ so that either $x - 1 = 4$ or $x - 1 = -4$. Hence either $x = 5$ or $x = -3$. But there are quadratic equations for which this method of "completing the square" fails. For instance, if we want to solve the equation $x^2 - 2x + 17 = 0$, we are led to finding a number x such that $(x - 1)^2 = -16$, and this seems impossible. The square of a number is never negative. As a matter of fact, there was no need to consider a somewhat elaborate example. The simple quadratic equation $x^2 = -1$ clearly has no solutions.

During the Renaissance an Italian mathematician had the brilliant and somewhat crazy idea of imagining what would have happened *if* there were a number whose square is -1. This number is denoted today by i and is called the "imaginary" unit. The agreement, originated during the Renaissance and still used today, is that we compute with i as if it were an ordinary number, but whenever the term $i^2 = i \times i$ appears, it is replaced by -1.

The equation considered before ($x^2 - 2x + 17 = 0$) now turns out to have solutions, namely $x = 1 + 4i$ and $x = 1 - 4i$. To check this, we compute that

$$\begin{aligned}(1 + 4i)^2 - 2(1 + 4i) + 17 &= (1 + 4i)(1 + 4i) - 2(1 + 4i) + 17 \\ &= 1 + 4i + 4i + 16i^2 - 2 - 8i + 17 \\ &= 1 - 16 - 2 + 17 = 0.\end{aligned}$$

But what good does it do to have a "solution" of a quadratic equation given by $x = 1 + 4i$? After all, i is an imaginary number, a mere plaything of the mathematician's imagination.

The Italian Renaissance mathematician Cardano (considered the father of imaginary numbers) noted, however, that in solving cubic equations (for instance, the equation $x^3 - 12x + 16 = 0$) it may happen that a calculation using imaginary numbers leads to roots that are all ordinary or "real" numbers. This observation showed that, imaginary or not, the square root of -1 is more than a toy.

Only in the nineteenth century did mathematicians succeed in removing the mystery from complex numbers. They did this by interpreting numbers like $2 + 3i$ or $\frac{1}{2} - 2i$ as names of points. More precisely, let us draw two perpendicular lines in the plane, one horizontally and one vertically, and choose a unit of length. We represent real numbers like 2 or $-\frac{3}{2}$ by points on the horizontal line: 3 is the point reached by moving 3 units to the right from 0, the intersection point of the two axes; $-\frac{3}{2}$ is the point reached by moving 3 half-units to the left. The imaginary unit i is represented by the point on the vertical axis one unit above 0; $-i$ is represented by the point on the vertical axis one unit below 0. A complex number like $3 + 2i$ is represented by the point in the plane reached by moving three units to the right and two units up from 0. (See Figure 1.) Thus,

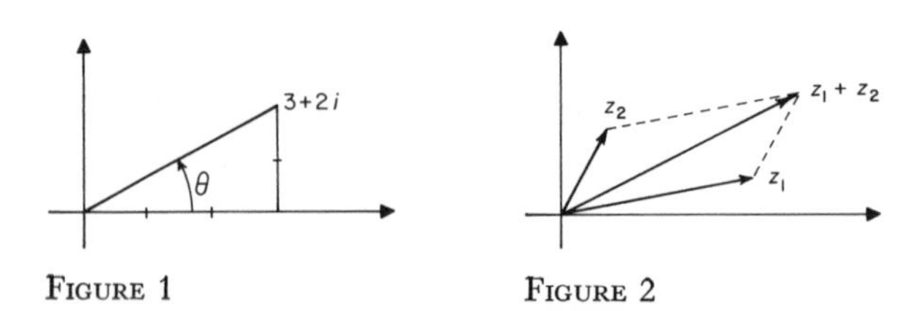

FIGURE 1 FIGURE 2

every point in the plane represents a complex number. Conversely, every complex number is represented by a point in the plane.

Nothing would be gained by representing numbers by points if we were not able to add and multiply them. The addition of two points is accomplished by the "parallelogram of forces," shown in Figure 2. It corresponds to the arithmetic definition of addition referred to earlier. The geometric definition of multiplication of two points is a little more involved. To multiply two points, z_1 and z_2 in Figure 3, we measure their distances r_1 and r_2 from 0 and also the angles θ_1 and θ_2 between the positive real axis and the lines connecting z_1 and z_2 to 0. We then locate the point z, whose distance from 0 is $r = r_1 \cdot r_2$ and whose "polar angle" is $\theta_1 + \theta_2$. This point or, rather, the complex number attached to it is called the

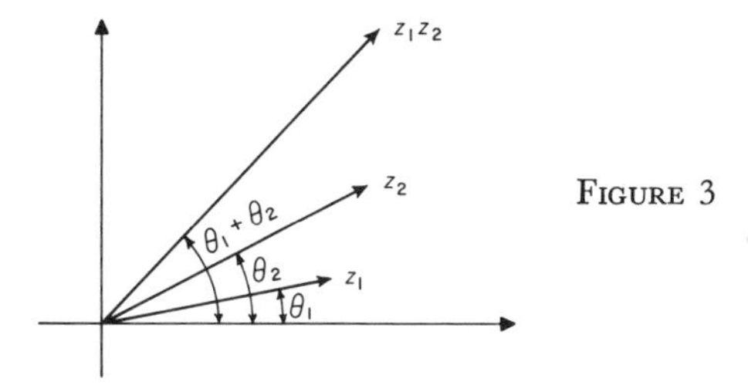

FIGURE 3

product of the points or complex numbers z_1 and z_2. Again the definition agrees with the arithmetic one. Let us verify this for the imaginary unit i. The point i has a distance 1 from 0, and its polar angle is 90°. The point $i^2 = i \cdot i$ therefore will have as distance from 0 the number $1 \cdot 1 = 1$ and will have the polar angle $90° + 90° = 180°$. It is therefore the point -1. Thus we have a geometric demonstration that $i^2 = -1$. We mention in this connection that the distance of a complex number $z = x + iy$ from 0 is called the modulus of z and is denoted by $|z|$. By the Pythagorean theorem, $|z|^2 = x^2 + y^2$.

Complex numbers were initially introduced to solve quadratic equations. One might have expected that, in order to solve every cubic equation, we would have to define some further kind of number, that equations of the fourth degree would require an even further extension of the number concept, and so on. Fortunately, this is not necessary. Once we have at our disposal complex numbers, every algebraic equation, no matter what its degree, can be solved. This principle holds even for such cumbersome oddities as $x^{100} - 2x^{73} + 10x^5 - 16 = 0$.

Complex Functions

A function, in mathematics, is a rule which associates with one number another number. In the simplest cases, this rule is given by an algebraic formula like $f(x) = 2x^3 - 6$. In words, this rule says, "Associate with every number x the number obtained by first multiplying x by itself three times, then multiplying the resulting number by 2, and then subtracting 6." Such a function, called a polynomial, is, of course, defined also for complex values of x. If x is $1 + i$, for instance, then

$$f(x) = 2(1 + i)\,(1 + i)\,(1 + i) - 6 = -10 + 4i.$$

But there are also functions defined in a quite different way. Important examples are the trigonometric functions sine and cosine. Suppose a point moves on a circle of radius 1 in a counterclockwise direction, beginning with a point A (see Figure 4). If the point has covered the distance x around the circle and is now at B, then $\sin x$ is the distance from B to the horizontal axis (counted as negative if B is below the axis) and $\cos x$ is

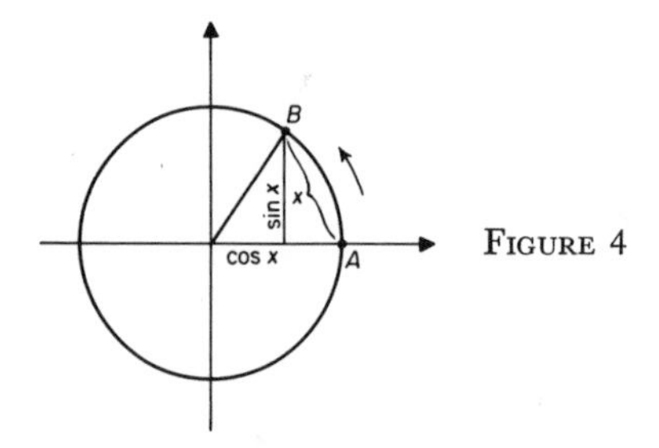

FIGURE 4

the distance from B to the vertical axis (counted as negative if B is to the left of that axis). The length of the whole circle is 2π. Therefore, after our point covers the distance 2π around the circle, it is again at A. This shows that $\sin x$ and $\cos x$ are periodic functions — that is, their values repeat every time the point moves around the circle. Thus, $\sin(x + 2\pi) = \sin x$, and $\cos(x + 2\pi) = \cos x$. Trigonometric functions were originally introduced and tabulated for calculations in astronomy and navigation and were used extensively in surveying. What is more important, they occur whenever we want to describe a periodic process like the vibration of a string, the tides, planetary motion, alternating currents, or the emission of light by atoms. The graphs of the sine and cosine functions are shown in Figure 5.

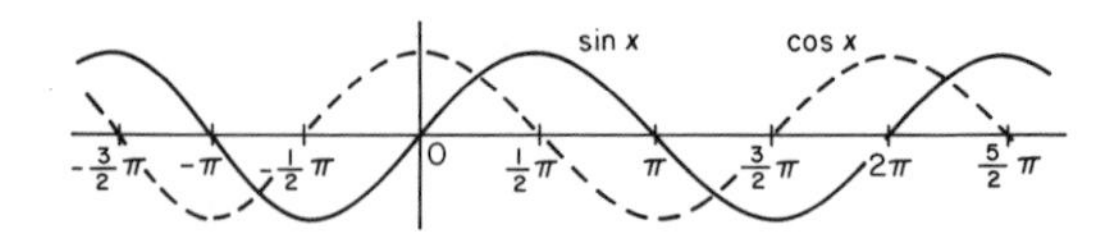

FIGURE 5

Another important function is the exponential function e^x. The number e (which is approximately equal to 2.718) may be defined as the amount obtained by depositing \$1 for one year in a fantastic bank that pays 100% interest and compounds it continuously. Similarly, e^x is the amount on deposit after x years. If x is an integer or a common fraction, the symbol e^x has the usual meaning. For instance, $e^2 = e \cdot e$, $e^{1/2} = \sqrt{e}$, $e^{-3} = 1/e^3$. The exponential function is used whenever we want to describe a process similar to the continuous compounding of interest, for instance, radioactive decay or the growth of a population. The graph of e^x is shown in Figure 6.

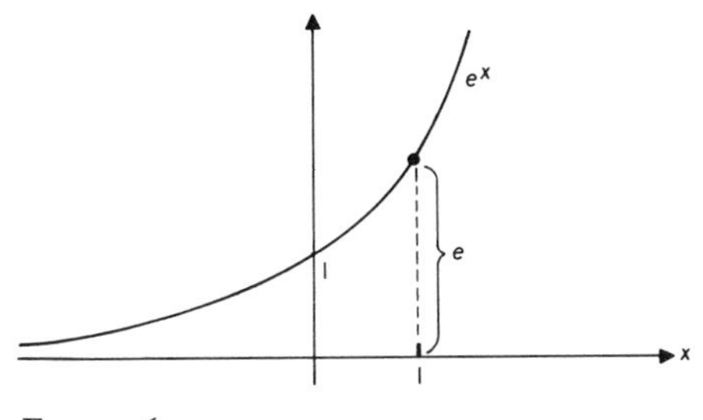

FIGURE 6

An important discovery in the eighteenth century, and one of the first applications of calculus, was the possibility of representing the exponential and trigonometric functions by "polynomials of infinite degree," or, as we now say, by "infinite series." The formulas are simple. First

$$e^x = 1 + \frac{x}{1} + \frac{x^2}{1 \cdot 2} + \frac{x^3}{1 \cdot 2 \cdot 3} + \frac{x^4}{1 \cdot 2 \cdot 3 \cdot 4} + \frac{x^5}{1 \cdot 2 \cdot 3 \cdot 4 \cdot 5} + \cdots .$$

The meaning of this is as follows: in order to compute e^x, we choose a whole number N and compute the value of the polynomial

$$1 + \frac{x}{1} + \frac{x^2}{1 \cdot 2} + \cdots + \frac{x^N}{1 \cdot 2 \cdots N},$$

that is, the sum of the first $N + 1$ terms of the preceding infinite series. If N is large, this value is close to that of e^x, and so we can compute e^x with any desired degree of accuracy by picking N large enough. Similarly it turns out that

$$\sin x = x - \frac{x^3}{1 \cdot 2 \cdot 3} + \frac{x^5}{1 \cdot 2 \cdot 3 \cdot 4 \cdot 5} + \frac{x^7}{1 \cdot 2 \cdot 3 \cdot 4 \cdot 5 \cdot 6 \cdot 7} + \cdots ,$$

and

$$\cos x = 1 - \frac{x^2}{1 \cdot 2} + \frac{x^4}{1 \cdot 2 \cdot 3 \cdot 4} - \frac{x^6}{1 \cdot 2 \cdot 3 \cdot 4 \cdot 5 \cdot 6} + \cdots .$$

These formulas have an unexpected consequence. Using them, we can compute e^x, $\sin x$, and $\cos x$ also for the case where x is a *complex* number. For instance, what is e^i? Is it e multiplied by itself $\sqrt{-1}$ times? Clearly this makes no sense whatsoever, but we can compute e^i by using the infinite series already written. The result is that e^i is a definite complex number: $e^i = 0.54030 + 0.84147i$, approximately.

Once the exponential and trigonometric functions are considered for complex values in this way, there exists between them a remarkable connection which reflects the resemblance among the three preceding infinite series. Specifically, it is a formula for e^{iy} in terms of $\cos y$ and $\sin y$:

$$\cos y = 1 - \frac{y^2}{1 \cdot 2} + \frac{y^4}{1 \cdot 2 \cdot 3 \cdot 4} - \frac{y^6}{1 \cdot 2 \cdot 3 \cdot 4 \cdot 5 \cdot 6} + \cdots,$$

$$i \sin y = +\frac{iy}{1} - \frac{iy^3}{1 \cdot 2 \cdot 3} + \frac{iy^5}{1 \cdot 2 \cdot 3 \cdot 4 \cdot 5} + \cdots,$$

$$e^{iy} = 1 + \frac{iy}{1} + \frac{i^2y^2}{1 \cdot 2} + \frac{i^3y^3}{1 \cdot 2 \cdot 2} + \frac{i^4y^4}{1 \cdot 2 \cdot 3 \cdot 4} + \frac{i^5y^5}{1 \cdot 2 \cdot 3 \cdot 4 \cdot 5} + \cdots.$$

Remembering that $i^2 = -1$ and $i^4 = 1$, we see that the third series is equal to the sum of the first two. Therefore,

$$e^{iy} = \cos y + i \sin y.$$

In particular, when $y = \pi$, we have $\cos y = -1$ and $\sin y = 0$. For this value, the formula becomes

$$e^{\pi i} = -1,$$

a beautiful relation between the four most important numbers in mathematics: e, π, i, and 1. Thus, the introduction of complex numbers discloses unexpected connections and harmony between seemingly unrelated parts of mathematics.

A function that can be expressed by a "polynomial of infinite degree" or a "power series" is called an "analytic" function. Every such function can be computed for complex values. It is fortunate that most functions of interest in pure mathematics and in applications of mathematics to the natural sciences are analytic. Complex analysis, the study of analytic functions, is therefore a central part of mathematics.

A Famous Function

Let z be a large positive number. Then the infinite series (not a power series)

$$1 + \frac{1}{2^z} + \frac{1}{3^z} + \cdots$$

can be summed. This means that there is a number, we call it $\zeta(z)$, such that, if we compute the finite sum

$$1 + \frac{1}{2^z} + \frac{1}{3^z} + \cdots \frac{1}{4^z} + \frac{1}{5^z} \cdots + \frac{1}{N^z}$$

for a large N, we get a value very close to $\zeta(z)$. The rule that assigns to a number z the value $\zeta(z)$ is a function; it is called the "zeta function" of Riemann.

There is another way of computing the same function. To describe it, we recall what is meant by a prime number. It is a positive whole number, not 1, which is not divisible by any number except 1 and itself. For instance, 2, 3, 5, 7, 11, 13, 17, 19, 23, and so on, are primes while 4 is not, being divisible by 2. The other way of computing $\zeta(z)$ is by the "infinite product" extended over all the primes 2, 3, 5, $\cdots$.

$$\zeta(z) = \frac{1}{1 - \frac{1}{2^z}} \cdot \frac{1}{1 - \frac{1}{3^z}} \cdot \frac{1}{1 - \frac{1}{5^z}} \cdots.$$

The fact that the infinite sum and the infinite product give the same value is, by the way, a consequence of the so-called fundamental theorem of arithmetic. This theorem says that every whole number can be written as a product of primes and, except for the order of the factors, in one way only. (For instance $12 = 2 \times 2 \times 3$, $162 = 2 \times 3 \times 3 \times 3 \times 3$.)

The reader who is not a mathematician and who can convince himself that the fundamental theorem of arithmetic implies that the infinite series and infinite product written above give the same value missed his calling. He should have become a mathematician.

By the formulas written above, the zeta function is defined for large positive values of z. Riemann, one of the greatest mathematicians of the nineteenth century, showed that there is a sensible way in which the zeta function can be defined for all real and complex values of z except for the single value $z = 1$. The zeta function is, of course, an analytic function. Riemann also realized that the study of the zeta function for complex values of z is a way of obtaining information about prime numbers. Many years after Riemann, toward the very end of the century, Hadamard (French) and de la Valée-Poussin (Belgian) used the zeta function and the whole apparatus of complex function theory developed partly by them to prove the so-called prime number theorem. This theorem asserts that the number of primes less than a given number x is approximately equal to $x/\log x$ or, what amounts to the same thing, to

$$\frac{x}{1 + \frac{1}{2} + \frac{1}{3} + \cdots + (1/x)}.$$

The words "approximately equal" have a very precise meaning. Let us denote the number of primes not exceeding x by the symbol $\pi(x)$. The prime number theorem asserts that

$$\frac{\pi(x)}{x/\log x} = 1 + \epsilon(x),$$

where $\epsilon(x)$ is a number that will be as small as we like if x is big enough. The theorem was first conjectured by Gauss from inspecting tables of primes.

It is certainly surprising and somewhat mysterious that a theorem about prime numbers was first proved by using complex analysis. The problem of finding a proof *not* involving complex analysis was solved only about eighteen years ago by A. Selberg (now at Princeton) and P. Erdös (a Hungarian residing in the United States at that time). Nevertheless, complex analysis remains the most powerful tool for obtaining information about prime numbers.

Though familiar to all mathematicians, Riemann's zeta function is far from a closed book. A great deal of research is focused on the nature of its roots. A root of a function is a number to which the function assigns the value zero. For instance, the roots of the function $\sin x$ are all the whole-number multiples of π. That is, $\sin x = 0$ when x is $0, \pi, -\pi, 2\pi, -2\pi, \cdots$. About the zeta function it is known that (1) every even negative integer is a root, (2) there are infinitely many other roots, and (3) all of the other roots are complex and have the form $s + ti$ where s and t are real and s lies between 0 and 1. Riemann himself conjectured that for all complex roots $s = \frac{1}{2}$.

To prove or disprove this Riemann hypothesis is universally considered the most challenging and difficult problem in today's mathematics. A proof of Riemann's hypothesis would give precise information about the error $\epsilon(x)$ in the prime number theorem and would at once establish a large number of other results in number theory. But the real reason for the awe with which all mathematicians regard Riemann's hypothesis is their knowledge that the greatest mathematicians since Riemann have tried and failed to solve this problem and that it has stimulated many profound and beautiful investigations.

Can we attack the problem by use of computing machines? Only in the sense that we can compute, with as high a degree of accuracy as desired, a finite number of zeros of the zeta function. No calculations have ever contradicted Riemann's hypothesis. If a single root should ever be found that contradicts the hypothesis, it would thereby be disproved. But no amount of calculation can ever establish the validity of the conjecture. If it is to be established at all, it will be by an act of thought and, unless we are all exceedingly mistaken, by an act of thought of the highest quality.

Conformal Mapping

The application to number theory is only one of the many interconnections between complex functions and other fields. Complex function theory also has a geometric aspect. This was first discovered by Gauss when he was planning a geodetic survey of a small German principality. Since the earth is not flat, no geographic map of a significantly large portion of the earth's surface can reproduce the correct ratios of size. Everybody has seen maps on which Greenland looks bigger than North America. On the other hand, it is essential for navigation that geographic maps should be, as they say, "conformal." This means that every angle, say the angle between a railroad track and a highway or a river, should be the same on the map as in reality. Early map makers, like Mercator and others, had learned how to make conformal maps of portions of the earth's surface. Gauss, being a mathematician, asked himself the general question: Is it possible to make a conformal map of any curved surface? He proved that it is possible for a small piece of the surface. He also recognized that, having one conformal map of a piece of a curved surface, one can obtain all other such maps by using complex function theory.

Let us explain this in more detail with an example. Consider a region R in the plane. A mapping of this region into another region R' is a rule which assigns to R a point in R' (Figure 7). We are interested in the case where every point in R' comes from some point in R and no two points in R are mapped into the same point in R'. We also assume that any two points in R close to each other are mapped into points in R' close to each other, and vice versa. Consider two intersecting curves A and B in R (Figure 8). Our mapping takes them into two curves A' and B' in R'. The mapping is called conformal if for any two such curves A and B the curves A' and B' always intersect at the same angle as A and B.

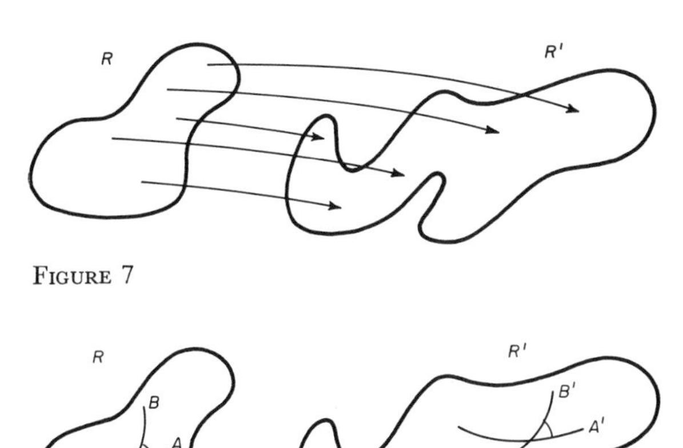

FIGURE 7

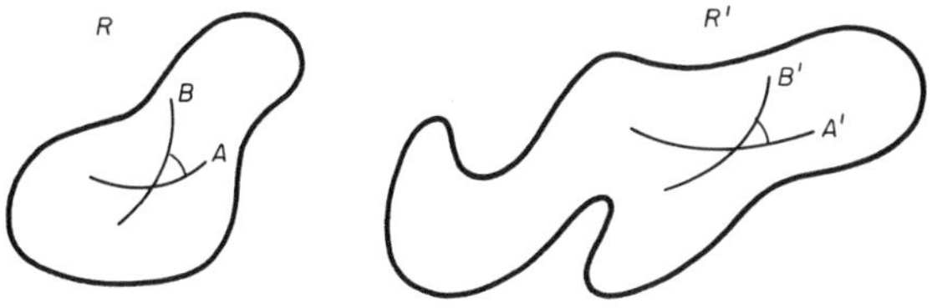

FIGURE 8

Now we noted earlier that points in the plane represent complex numbers. Our mapping, therefore, is a rule that assigns to every complex number represented by a point in R another complex number represented by the corresponding point in R'. The mapping is described by a function of a complex variable. And it turns out that to say that the mapping is conformal amounts to saying that the function describing it is an analytic function (more precisely, an analytic function with a nonvanishing derivative). Some examples of conformal mappings given by simple analytic functions are shown in Figures 9 and 10.

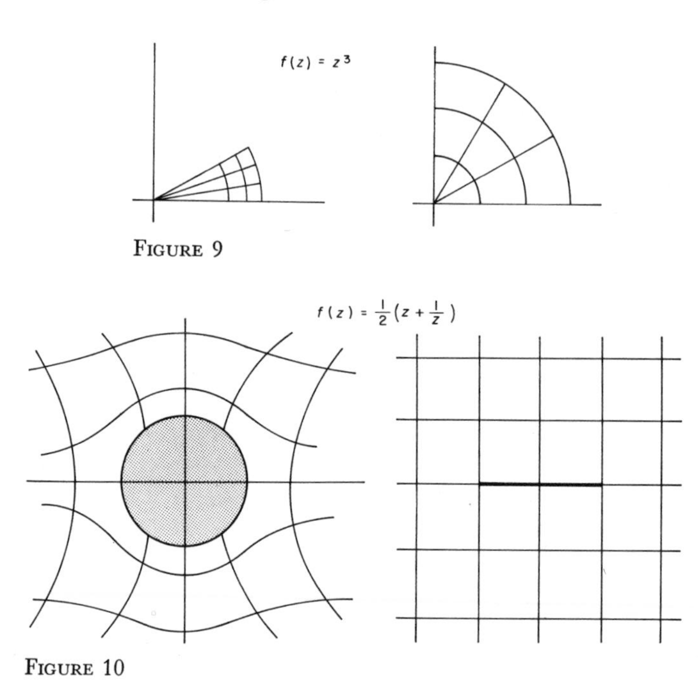

FIGURE 9

FIGURE 10

The theory of conformal mapping is one of the most highly developed parts of complex function theory. A central result is Riemann's mapping theorem, which says that every region bounded by a single curve can be mapped conformally onto any other such region. For example, a square can be mapped conformally onto a circular disc (see Figure 11). Riemann's proof of the mapping theorem was magnificently simple, but it contained a basic mistake, an almost unheard-of thing in mathematics.

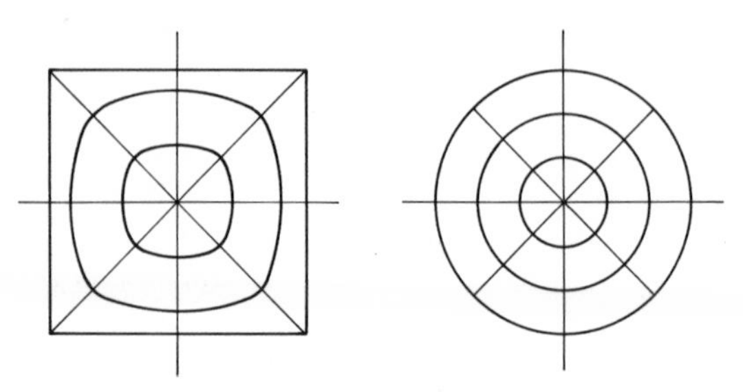

FIGURE 11

For many years a rigorous proof of the mapping theorem was a challenging problem for mathematicians, and several flourishing mathematical disciplines owe their origin to this problem. Many proofs are now available. Surprisingly, it turns out that Riemann's original approach can be carried out correctly.

Let us now turn to an unsolved problem connected with conformal mapping in order to give the reader an idea of the flavor of modern mathematical research. If we consider a conformal mapping of the disc of radius 1 centered at the origin (the zero point) onto some other domain, then the mapping is described by an analytic function, $f(z)$, which can be expressed by an infinite power series

$$a_0 + a_1 z + a_2 z^2 + a_3 z^3 + a_4 z^4 + a_5 z^5 + \cdots .$$

A well-known conjecture, usually associated with the name of Bieberbach, states that $|a_n|$ is not greater than $n|a_1|$. Note that this contains infinitely many statements, one for each index $n = 2, 3, 4, \cdots$. The statement for $n = 2$ is easy and has been proved by Bieberbach. The case $n = 3$ is incomparably more difficult and was established by Loewner (later at Stanford) in 1923. The next step was taken thirty years later. Schiffer and Garabedian (at that time both at Stanford) proved in 1953 the Bieberbach conjecture for $n = 4$. The general case where n can be any integer at all is still an open question, though there are many partial results pointing toward the truth of the conjecture. It is known, for instance, that the Bieberbach conjecture is true if the function maps the unit disc onto a convex region or onto a symmetric one. It is also known that, for every given function, the Bieberbach conjecture can be false for at most finitely many coefficients (Hayman, England, 1954). Although the Bieberbach conjecture does not have the status of the Riemann hypothesis, its challenge is felt very keenly by many excellent mathematicians, and the mood at the moment is quite optimistic. The solution of the problem will hardly produce results of interest to other questions, but it seems likely that the methods developed will be very useful elsewhere in pure and applied mathematics. This applies in particular to the attempt to use electronic computers as a heuristic device (Garabedian). This means that the computer is not supposed to prove a theorem but, rather, to show the mathematician which of the many conceivable approaches is promising.

Applications of Conformal Mapping

Conformal mappings are widely useful in problems of mathematical physics. We mention only one application to fluid dynamics. Consider, for instance, a conformal mapping of a straight strip, as shown in Figure 12, onto a channel. The images of the horizontal lines are then streamlines of a possible flow of a fluid, and the images of the vertical lines are the

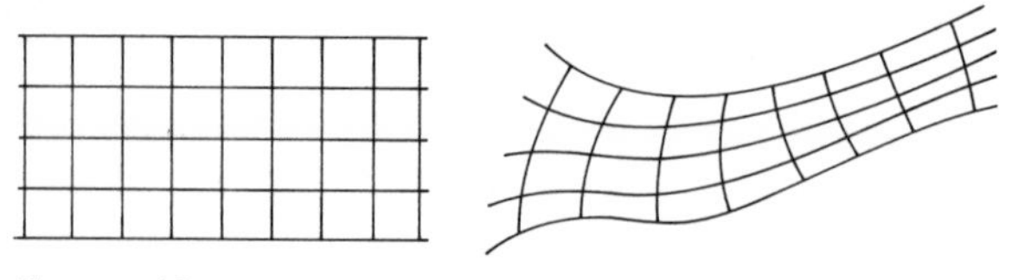

FIGURE 12

so-called equipotential lines. The sizes of the images of the little squares permits us to compute the pressure on the walls exerted by the fluid.

It is possible, however, to describe the fluid flow in this way only under very restrictive assumptions. It must be a steady, two-dimensional, irrotational flow of an incompressible, frictionless, and weightless fluid. While such fluids do not exist in nature, there are, nevertheless, cases in which a calculation based on complex function theory gives a usefully accurate description of an actual physical process.

This applies, within certain limits of course, to the flow of air around an airplane wing. Since the pioneering work of Kutta and Joukowiski in the beginning of the century and extending to the end of World War II, the design of airplane wings was based on a theory involving the conformal mapping of a cross section of the wing, the so-called "profile," onto a circle (Figure 13).

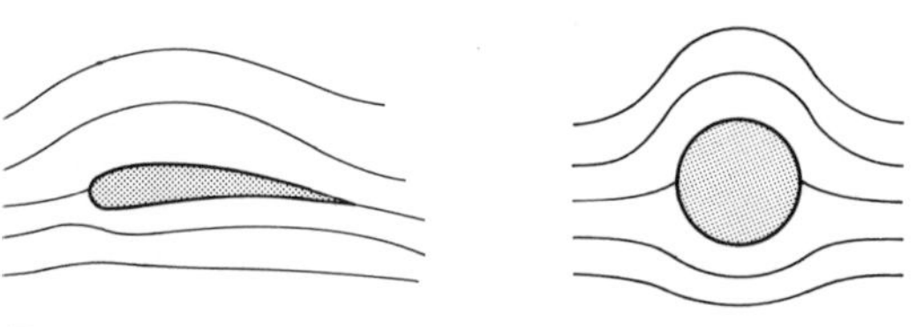

FIGURE 13

The theoretical pressure distribution on an airplane wing computed by conformal mapping did no more than provide guidelines for experimental work and for actual design. Now that the development is essentially complete, it would be of very great interest to investigate in detail how and through which mechanisms mathematicians, theoretical fluid dynamicists, experimental fluid dynamicists, and airplane designers actually communicated among themselves and influenced each other's work. With the ample data now available, an investigator with both scientific and historical training could perform a useful service.

Although actual airplane designers hardly ever performed conformal mappings, the problem of computing a conformal mapping of a given profile attracted much attention. Very effective methods were developed even prior to the advent of modern computers, and with computers available the problem is, for practical purposes, solved.

It would be tempting to rewrite history and claim that the discovery of the basic theorems in function theory, such as Riemann's mapping theorem, was motivated by the needs of fluid dynamics. But this is not so. It is likely that Riemann was influenced in his thinking about complex functions by physical imagery, but he proved the mapping theorem while working on the theory of algebraic functions and their integrals, a theory that is, thus far, of interest in pure mathematics only. The usefulness of the mapping theorem to applications came many years later.

The attempt to extend the theory to high-speed subsonic flows led to generalizations of complex function theory. These proved to have little or no value for applications since it turned out that relatively simple rules of thumb were sufficient for all practical purposes. However, the generalizations did prove to be of value in pure mathematics.

Some Recent Developments

Among the many aspects of complex function theory now being investigated, none is more fascinating than the theory of functions of several complex variables. At first glance it may seem that no essentially new phenomena will be encountered if, instead of studying functions of one variable, we consider power series in two or more variables. Indeed, the simplest functions considered in analysis lead to functions of more than one variable. For instance, sin (ax) may be considered as a function of the two numbers a and x; both may take on complex values; and the resulting function is analytic.

Analytic functions of one variable are considered as defined in parts of the plane because complex numbers may be thought of as represented by points in the plane. In certain more-advanced chapters of classical function theory, we consider also functions defined on other two-dimensional surfaces. Functions of several complex variables, however, are always defined in regions of a space of many dimensions, at least four. This not only exceeds the powers of our visualization but also leads to completely new phenomena. Here is a simple example. The simplest functions, next to polynomials, are the so-called rational functions. These are quotients of two polynomials. A rational function of one complex variable can be written as

$$r(z) = p(z)/q(z)$$

where p and q are polynomials that we assume have no common roots. The denominator will equal 0 at certain points, the roots of q. These points are called the poles of the rational function. If z is close to a pole, the function takes on large values. We say that at a pole the function becomes infinite. The situation with rational functions of several variables is quite different. For instance, the function

$$r(z_1, z_2) = \frac{z_1}{z_2}$$

is said to have a pole at every point (z_1, z_2) where $z_1 \neq 0$ and $z_2 = 0$. At such points r becomes infinite. But if the point (z_1, z_2) approaches the point $(0, 0)$, r does not approach any definite limit. This point is called a point of indetermination of the function. For functions of more than two variables, points of indetermination occur not singly but on what are called higher-dimensional surfaces.

Another peculiarity of functions of several complex variables had been observed by Poincaré. In one variable there is no essential difference between functions defined in a disc or in any other domain bounded by a single curve. This is a consequence of Riemann's mapping theorem. In several variables, however, domains that look very much alike from the viewpoint of geometry may be completely different from the viewpoint of analytic functions that "inhabit" them. For this reason, the modern theory of analytic functions of several variables has a strong geometric flavor, and its methods and results are intimately connected with such fields of mathematics as topology, differential geometry, and algebraic geometry. Some of the most striking and penetrating results of this theory were obtained over a period of several decades by the contemporary Japanese mathematician, K. Oka, who has devoted his scientific life to this discipline. But the reformulation and extension of Oka's ideas in the language of modern mathematics, a task performed primarily by French mathematicians, played a big part in the growth of the theory.

Classical function theory was always interwoven with differential geometry on the one hand and the theory of partial differential equations on the other. The theory of functions of several complex variables has even more connections with other parts of mathematics. Those who believe that the harmony between mathematics and the natural sciences is predestined will not be surprised to learn that the theory of several complex variables is already being used in modern physics (see Wightman's essay in this collection).

BIBLIOGRAPHY

Z. Nehari, *Introduction to Complex Analysis*, Allyn & Bacon, Boston, 1961.
Z. Nehari, *Conformal Mapping*, McGraw-Hill, New York, 1952.
L. V. Ahlfors, *Complex Analysis*, McGraw-Hill, New York, 1966.

BIOGRAPHICAL NOTE

Lipman Bers was born in Latvia and came to the United States soon after the outbreak of World War II. He participated in the Research and Training Program in Applied Mathematics at Brown University and later taught at Syracuse University and at New York University. He is now a professor at Columbia University. His research on partial differential equations and complex analysis stemmed from wartime work on the flow of compressible fluids, which led rather naturally to generalizations of complex function theory. This research led Bers to apply generalized complex functions to classical function theory. Thus, as he has remarked, he reversed the usual pattern of a mathematician's career by proceeding from applications to pure mathematics and from generalizations to specifics.

Although social scientists still describe more than they measure, they are becoming dependent on mathematics. Some of the techniques they find most useful have been developed in recent years, and in the future their problems will very likely inspire much new mathematics. Mathematically speaking, the social sciences are likely to be far more difficult than physics. This should not be surprising since the negotiations of a committee are far more complex than the orbits of a solar system. Moreover, social scientists often need mathematics that is geared to qualities rather than numbers. This essay, through several highly simplified examples, indicates how social scientists use mathematics even when they are unable to make precise measurements. The topics range from a typical business decision to a natural balance between foxes and rabbits in a forest. The early successes of mathematics in the social sciences suggest that the best is yet to come.

The Social Sciences Call on Mathematics

John G. Kemeny

There is a long tradition of cooperation between mathematics and the physical sciences. The latter call on the former for the solution of theoretical problems, while the former draws inspiration from these problems. Much first-rate mathematics has its roots in applications.

The connection between mathematics and the social sciences is a more recent development, and it is still in an embryonic stage. But during the past two decades a number of highly promising new applications of mathematics have arisen. It is the purpose of this essay to illustrate this development. The examples will, necessarily, be oversimplified, but they may convey some of the flavor of the new developments.

The importance of statistics has long been recognized in the social sciences. Therefore, we have chosen nonstatistical applications. The examples have been chosen to illustrate both the variety of mathematical techniques now in use in the social sciences and the diversity of subject matter that uses mathematics.

We may expect that, as the social sciences develop, they will become more mathematical. This is an invariable feature of the maturing of any science. As the subject grows, it becomes more precise and requires an apparatus for accurate predictions. This apparatus is usually based on a branch of mathematics. Common objections to this position are that many social sciences are nonquantitative or that they have a strongly random element. But modern mathematics is prepared to deal with "qualitative" theories, and it has powerful tools for studying chance events.

The real difficulty in applying mathematics lies in the fact that the social sciences are more complex than the physical sciences. The behavior of a committee of human beings is vastly more complex than the orbits of a planetary system. This accounts for both the late development of the social sciences and the difficulty in employing mathematical methods. But in time we may expect mathematical tools of high sophistication. Perhaps the most interesting question for the mathematician is whether present-day mathematics is adequate for coping with so complex a subject matter. If not, then the social sciences may serve in the future as a source of inspiration for new branches of mathematics, as the physical sciences have served in the past.

The reader interested in some of the methodological problems is referred to [1] and [4] in the references.

Linear Programming and the Theory of Games

Two important new branches of mathematics, both motivated by economic problems, were developed in the 1940's: linear programming and the theory of games. While these subjects were independently developed, it was later shown that there is a strong mathematical connection between them, from which both subjects profited. Linear programming is simpler, and we shall concentrate most of our attention on it. We shall return to the theory of games briefly at the end of the section. An elementary introduction to both fields may be found in [5, Chapter 6].

A typical linear programming problem is one faced by a manufacturer who has several methods of production (or marketing) at his disposal and wants to use his resources so as to maximize his profits. Another type of application is one in which the businessman tries to achieve a given objective at a minimum cost.

It is easiest to describe the method in terms of a concrete example — and we select a very simple one. The Tasty Nut Company packages salted nuts for retail sale. The company has on hand 84 pounds of peanuts, 40 pounds of cashews, and 27 pounds of walnuts. It packages two mixes. The "regular mix" consists of 70 percent peanuts, 20 percent cashews, and 10 percent walnuts. The "fancy mix" contains 40 percent cashews and 30 percent of each of the other kinds. The regular mix sells for $1.20 per pound and the fancy for $2.00. The objective is to mix the nuts so as to maximize total receipts, or profit.

Our first task is to describe what courses are open to the firm. Let x be the number of pounds of regular mix packaged, and y the number of pounds of fancy mix. Clearly,

$$x \geq 0 \qquad \text{and} \qquad y \geq 0. \tag{1}$$

We must also make sure that the company does not attempt to use more of a given kind of nuts than it has on hand. For example, x lb of regular mix require $0.7x$ lb of peanuts, and y lb of the fancy mix uses up $0.3y$ lb of peanuts. Thus we require that

$$0.7x + 0.3y \leq 84. \tag{2}$$

Similarly, for cashews and walnuts, respectively,

$$0.2x + 0.4y \leq 40. \tag{3}$$

$$0.1x + 0.3y \leq 27. \tag{4}$$

These conditions are shown in Figure 1. Condition 1 means that we must stay above the x-axis and to the right of the y-axis. Conditions 2 through 4 are represented by the three slanting lines, and in each case we

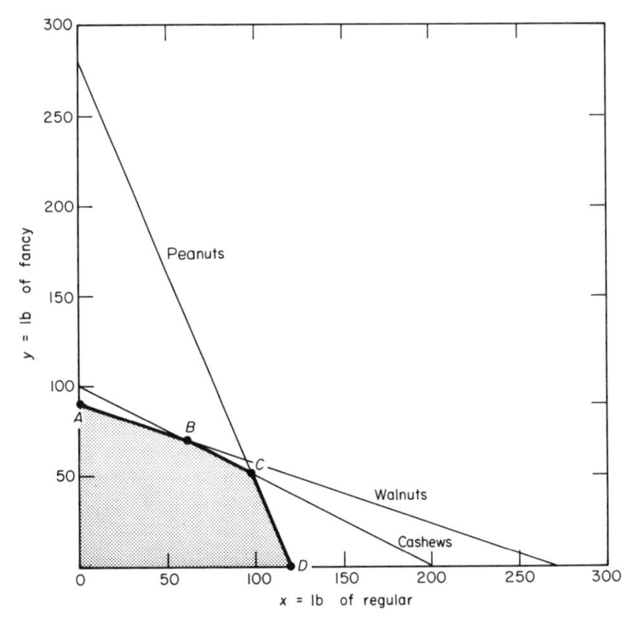

FIGURE 1

must stay below (or on) the line. This restricts the solution to the shaded five-sided region.

So much for the possibilities. What solution brings in the largest income? This will, of course, depend on the prices charged for the two mixes, but we can make some general observations. There is no sense in picking a solution in the interior of the shaded region since by moving "up" we will increase the profit. Thus, we may restrict our attention to the three heavily drawn line segments, which form the upper boundary. But this still allows infinitely many solutions in principle — and a very large number in practice. Here is where the fundamental theorem of linear programming comes to the rescue: A best solution is always found at a corner. Hence, we need only examine the four corners labeled A, B, C, and D.

We can find the coordinates of these corners by simple algebra:

$$A = (0,90); \quad B = (60,70); \quad C = (98.2,50.9); \quad D = (120,0).$$

Since the regular mixture brings in $1.20 per pound and the fancy mix $2 per pound, the income corresponding to the four corners is

$$A = \$180, \qquad B = \$212, \qquad C = \$219.64, \qquad D = \$144.$$

Thus, C is best; the company should package 98.2 lb of regular mix and 50.9 lb of the fancy mix. This uses up all of its peanuts and cashews, and nearly 2 pounds of walnuts are left over. This method will be successful as long as costs and gross receipts are "linear." For example, this assumes that doubling sales will double the gross receipts, which is presumably true in our example and is at least approximately true in many applications.

How well could the company have done by trial and error? They might have tried, for example, to package the same amount of both mixes. This would have produced 66 lb of each. Then they might have noticed that, although their supply of cashews and walnuts was almost exhausted, they could squeeze out 2 more pounds of the regular mix. This would have yielded them $213.60, a respectable income. But they would have been 3 percent short of the best they could do.

If prices change, it is not necessary to start from scratch. The conditions of Figure 1 are still applicable. One simply retests the corners. Let us say that the price of the regular mix has to be lowered. The same solution C is best as long as the regular mixture brings in at least $1.00; if the price falls below this, the company switches to solution B.

Although this example is certainly oversimplified, many important economic decisions can be treated by the same mathematical techniques. Suppose that a company manufactures 50 commodities, using 100 raw materials. This becomes a 50-dimensional problem, with 100 restrictions taking the place of the 3 slanted lines. We can no longer draw a diagram, but the principle is the same. The restrictions force us to stay within a complicated box in 50-dimensional space, and the fundamental theorem guarantees that we need consider only corners. Thus, we need a systematic method for finding corners and evaluating them until we find the solution yielding the largest profit. Such methods have been developed, and a high-speed computer can easily cope with the size problem just described. In this case, linear programming should yield a solution that is much more than 3 percent better than the best that the company would achieve by trial and error.

Most important is the fact that linear programming provides not only the best solution but also an absolute guarantee that the best solution has been found. This can save many worries for management.

These problems deal with a single manufacturer. What if two or more businesses are in competition? Then we turn to the theory of games. While the theory talks in terms of games of strategy, it is ideally suited for the analysis of many competitive situations. When two firms compete, the theory is as completely known as the theory for linear programming. But when two firms are able to cooperate or when there are more than two sides, the theory is as yet incomplete. Nevertheless, it provides a rigorous language for the formulation of the problem, and it provides partial solutions.

For an excellent discussion of both traditional theories and new theories of the market, formulated in terms of the theory of games, the reader is referred to Shubik's analysis in [7].

Graph Theory

A common objection to the use of mathematics in the social sciences is that the information available may be only qualitative, not quantitative. There are, however, several branches of mathematics that deal effectively with qualitative information. A very good example is the theory of graphs. The recent book [2] contains a good introduction to both the theory and application of graphs.

Let us consider a set (collection) of points, some of which may be connected by lines. We think of each line as having a direction, indicated by an arrow on the line. Thus, two points may not be connected at all, have a one-way connection, or have a two-way connection. Such a collection of points and directed lines is known as a *directed graph*. Examples of directed graphs are shown in Figure 2.

Figure 2a might represent part of a police communication network. Point 1 is headquarters, 2 and 3 are squad cars, while 4 and 5 are officers

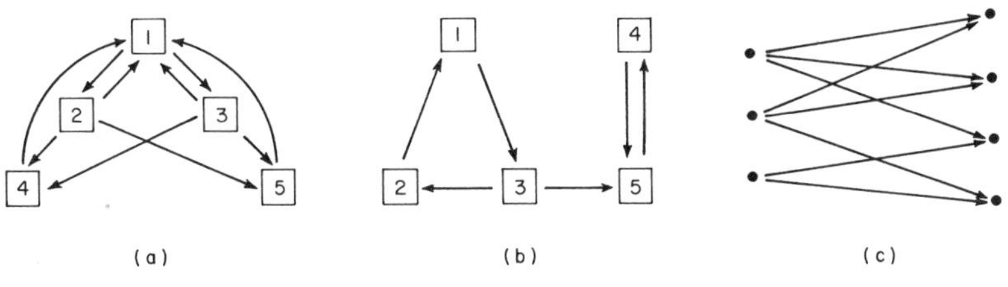

FIGURE 2

patrolling different parts of town. All three types of connection are illustrated. There is two-way communication between headquarters and the squad cars. There is one-way communication between a car and a patrolman since the car can find the patrolman and, similarly, the patrolman can call headquarters, but headquarters cannot call a patrolman directly. And there is no direct communication between the two cars or the two patrolmen.

This is an example of a *strongly connected* graph since everyone can send a message to everyone else, although sometimes only indirectly. For example, if a patrolman wants to make contact with a squad car, he telephones headquarters and asks them to radio the car. Figure 2b is connected but not strongly: 1 can send a message to anyone, but 5 cannot send to 1, even through an intermediary.

A large body of theorems exists for directed graphs, many of which are useful for applications, such as the study of communication networks. For example, [2, Chapter 7] deals with the vulnerability of communication nets to the removal of lines. It develops various measures of the importance of single lines and of the vulnerability of the entire network. This could be useful in deciding how best to disrupt enemy communications or to protect one's own system.

We will illustrate the theory of directed graphs by means of another problem. Several families are going on a picnic, using a variety of cars, motorcycles, and so on. To make sure that there are no fights, we insist that no two members of the same family share a vehicle. Therefore, the passengers of a given vehicle must all come from different families. Can we accomplish our goal?

Of course, we need to know how many members there are in the various families and how many seats there are in each conveyance. Let us say that the ith family has p_i members, and the jth conveyance has s_j seats. To make the problem as hard as possible, let us assume that there are barely enough seats. Mathematically this is stated as

$$\sum_i p_i = \sum_j s_j; \tag{5}$$

that is, the total number of people equals the total number of seats.

First we observe that the problem is not always solvable. For example, if the first family consists of 10 people, and there are only 8 conveyances, we cannot put them all in different conveyances, even if Condition 5 is satisfied. It is up to the mathematician to find a useful criterion that will assure that the problem has a solution.

We will next derive a set of conditions that are clearly necessary for a solution. Let us consider the first k families. How can we distribute them? Of course, car j can take at most s_j of them since this is the number of seats. But, since we cannot put two people from the same family into the same car, car j can take at most one from each family — or k people. Let

$\min(k, s_j)$ be the smaller of the numbers k and s_j. Then we can put at most $\min(k,s_j)$ people from the k families into the jth car. To assure that we can accommodate the k families, we must require

$$\sum_{i=1}^{k} p_i \leq \sum_{j} \min(k, s_j). \tag{6}$$

And this must be true for every k.

Let us illustrate these two conditions in terms of a concrete example. We have three small families, with 3, 3, and 2 members, respectively. And we have 4 two-seater sports cars. Then

$$p_1 = p_2 = 3, \qquad p_3 = 2, \qquad \text{and} \qquad s_1 = s_2 = s_3 = s_4 = 2.$$

Since $3 + 3 + 2 = 2 + 2 + 2 + 2$, Condition 5 is satisfied, and we have exactly the right number of seats. We note that $\min(1, s_j) = 1$, $\min(2, s_j) = 2$, and $\min(3, s_j) = 2$ for each car. Thus, writing the Conditions 6 for $k = 1, 2, 3$, we have

$$k = 1: \quad 3 \leq 1 + 1 + 1 + 1,$$
$$k = 2: \quad 3 + 3 \leq 2 + 2 + 2 + 2,$$
$$k = 3: \quad 3 + 3 + 2 \leq 2 + 2 + 2 + 2,$$

and all the conditions are satisfied.

What does this have to do with graph theory? We may think of the solution of this problem as represented by a directed graph. On the left of the graph we have points representing the families, on the right, points representing the cars. Each time a member of a family is assigned to a conveyance, we draw an arrow from the family's point to the car's point. Our problem then becomes one of drawing a specified number of lines starting at each family point, with a specified number ending at each car point. The question of whether such a graph exists is solved in [2, Chapter 12]. We learn that if the families are listed starting with the largest family, then the next largest, and so on, and if Conditions 5 and 6 are met, there always exists a solution. For example, the solution of the problem of the three small families and the four sports cars is shown in Figure 2c.

As another concrete example, consider five families with 5, 5, 5, 3, and 2 members, respectively. Let us suppose that each family provides its own car, which is just large enough for that family. We then have 20 people and 20 seats, so that Condition 5 is met. Condition 6 is all right for $k = 1$ and 2, but, for $k = 3$,

$$5 + 5 + 5 \leq 3 + 3 + 3 + 3 + 2$$

is false. Hence this problem has no solution. The reader is encouraged to convince himself of this fact by trying to draw a graph that assigns the people to cars.

The reader who is intrigued by the idea of "qualitative mathematics" will find four examples discussed in "Mathematics without numbers" in [4].

Differential Equations

The fact that most of the applications discussed in this essay use twentieth-century mathematics is not an accident. Some of the more recent advances in mathematics have proved to be most useful for the social sciences. Indeed, the social sciences have served and still do serve as motivation for the development of new mathematics.

But we want to show at least one application of classical mathematics. The theory of differential equations has many applications in economics. However, we have selected an amusing — and instructive — example from ecology, due to Lotka (see [6]). The treatment we give this model is taken from [3].

Let us consider two species of animals, such as foxes and rabbits, one of which feeds on the other. Given the natural growth rates of the species and the effect of the interaction, what can we say about the growth of the species?

Differential equations are ideally suited to describe the change of a given quantity as time passes. For example, if we let x stand for the number of rabbits at a given time and if there were no foxes to hunt them, rabbits would multiply in proportion to the number of rabbits. We express this by

$$\frac{dx}{dt} = Ax \tag{7}$$

or "the rate of change of rabbits is proportional to the number of rabbits." The number A measures how prolific the rabbits are.

On the other hand, if y stands for the number of foxes at a given time and if there are no rabbits to feed on, then the foxes would die out. We therefore write

$$\frac{dy}{dt} = -Py, \tag{8}$$

where P is the death rate, and the minus sign indicates that the number of foxes is decreasing.

We may assume that the number of kills of rabbits is proportional to xy, the product of the two numbers. Such kills decrease the rabbit population but help the fox population grow. Therefore, we add an xy term to each equation, with a proportionality constant and with a plus sign for foxes but a minus sign for rabbits. Our completed model now takes the form

$$\frac{dx}{dt} = Ax - Bxy, \tag{9}$$

$$\frac{dy}{dt} = Cxy - Py. \tag{10}$$

The numbers A, B, C, and P are all positive, and they must be established from ecological observations. But the interesting feature of this model is that the mathematician can tell the ecologist a great deal even without knowing the exact values of the four constants!

We can predict a cyclic change in the two populations, as shown in Figure 3. We have plotted the number of rabbits on the horizontal axis

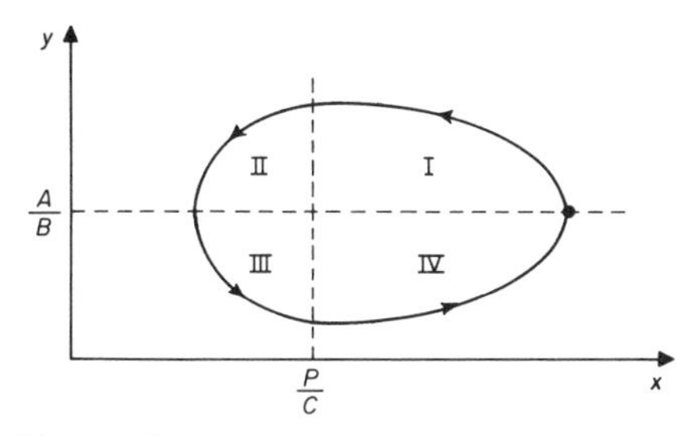

FIGURE 3

and the number of foxes vertically. Suppose that we start at the right-hand end point, marked with a heavy dot. At this stage, and throughout region I, both rabbits and foxes are plentiful. Thus, the foxes are on the increase; however, the rabbits are decreasing since there are too many foxes around. This is shown by movement "upward and to the left." When we reach the top of the curve, the number of rabbits has dropped to a critical value P/C, which means that the foxes no longer have sufficient food and both populations are on the decline through region II. At the left-hand end point, the number of foxes has dropped to A/B, which means that the rabbits can start to increase their number, and this happens through region III. Finally, when we reach region IV, foxes again start to increase until we get back to the starting point. Then nature repeats itself.

The values P/C and A/B are not only signals for a change in the pattern, they also turn out to be long-term averages for the number of rabbits and number of foxes, respectively. Thus, the species oscillate around these averages in cyclic fashion.

Mathematical theory reveals all this information without knowing the exact values of A, B, C, and P. If we know these values, we can, of course, determine the exact length of a cycle and predict the number of foxes and rabbits at each stage.

Computer Simulation

It is generally recognized that high-speed computers are valuable tools in the development of mathematical models. The normal procedure is somewhat as follows. The social scientist collects a large amount of data. This leads him to some general principles which, when formulated in a mathematical language, provide a theoretical model. It is the task of the mathematician to develop this formulation. Theoretical mathematics reduces problems to routine computations, and the computer is the tool for carrying out the computations.

The heart of this development is, indeed, the formulation of the model and its mathematical development. But the computer may be valuable in connecting theory to experience. However, there is one use of computers in which a computer program is not a tool but actually a substitute for a mathematical model. This occurs when a computer is used to *simulate*, or act out, a situation from real life.

In many situations the problem is too complex for theoretical treatment but the outcomes are fairly well understood. It may then be useful to ask a computer to act out the circumstances, so that we may study them under laboratory conditions. (See [4], "Games of Life and Death.")

Let us illustrate such a simulation model. The mathematical theory of queues (or waiting lines) is highly developed, but many actual queues are too complex for such treatment. A large department store may have hundreds of salesmen, with a wide variety of products, and customers requiring all kinds of service. The store can collect detailed statistics on the number of customers on a given day, their distribution by time of day and by what they are looking for, and how much attention a given type of customer requires. It wants to find out the minimum sales force that will assure reasonably good service and how the sales clerks should be scheduled and distributed. This problem is too complex for ordinary mathematical analysis, but it is ideal for machine simulation.

The planner supplies to the computer all the available customer information as well as the present distribution of the sales force. The computer then runs through a year's experience, noting where undue lines build up. The planner then modifies the distribution of his clerks. He may add to the force if service is too poor, or he may attempt to save by having a somewhat smaller force better distributed. Again, the machine runs through a year, noting the results. Thus, by trial and error, a near-ideal solution may be worked out.

This is trial-and-error learning. But the trials are carried out inside the memory of the computer and not at the expense of the business or the customers. The planner can also acquire years of experience in a few hours.

The same simulation model may be used to train a new floor manager or a new vice president. The computer acts out a typical month in the store, and the executive is called upon to make decisions as queues build up. His decisions are incorporated into the acting out. Such training models are now in use in several business schools. Students may acquire years of experience on the stock market by having the computer simulate several years of the history of the stock market and keeping track of how the students are doing in their investments. Or it may simulate several large companies fighting for a given market, with the students making managerial decisions for one of the companies.

Chance plays an important role in most social phenomena. Therefore, a realistic simulation should include a role for randomness. Modern computers are ideally suited for doing so. One can generate a sequence of "random numbers" that behave very much like random devices. For example, to imitate the roll of a die, one generates a sequence of random digits from 1 through 6. In a sequence of one million "rolls" the computer-simulated die will look more random to the statistician than a physical die rolled by a human being!

We shall describe one large simulation study in detail, taken from [5]. Although it was concerned with baseball, it illustrates many of the techniques used by social scientists in simulation models.

The problem was to determine the effect (if any) of the batting order on the production of runs in baseball. The study used data on the 1963 world champion Los Angeles Dodgers. The nine players of the starting line-up were introduced to the computer, with information on how they usually performed at bat. For each player, the computer was furnished the probabilities of an out, a walk, a single, a double or triple, and a home run. For the pitcher's spot, an "average pitcher" was used.

The computer was then taught the rules of baseball and rules for moving batters around the bases. The model was tried out by having the side bat through an entire season of 162 games. Detailed statistics were kept and compared with the 1963 season records of the Dodgers.

Because of the heavy influence of chance, the results did not agree completely with the actual season, but they differed from it no more than one season differs from another in reality. For example, the total number of runs scored in the simulated season was 652, compared with 640 in the actual season. The Dodgers' leading batter did even better on the computer while one of the weaker batters did much worse than in 1963. But, generally, batting averages were close to those of the actual season.

Of course, the model was incomplete. For example, it did not allow stealing of bases, or double plays, or pinch hitters, or extra innings. Therefore, certain statistics (like men left on base) turned out to be unrealistic. But as far as the problem studied was concerned, that is, the production of runs, the model was highly realistic.

Then a variety of batting orders were tried, first for 10 seasons each and later for 70 seasons. Only in the latter horrendously long computation did statistically significant differences turn up.

The conclusions drawn from this simulation are quite interesting. The batting order does make a difference. However, one has to do something pretty stupid before the difference is significant, such as putting several of the worst batters near the beginning of the line-up. And even the worst line-up will cost the team only about 30 runs per season. This is significant but surprisingly small. And the minor changes in order so dear to managers appear to make no difference at all! Pure luck is a much more significant factor; the changes from season to season for the same line-up are considerably larger than the differences in the average numbers of runs produced by slightly different batting orders.

This illustration shows up both the power of the method and its severe shortcomings. We have a method of attacking so complex a problem as the number of runs produced during a season by a given batting order (much too complex for theoretical treatment). And the simulation can quickly give us a feeling for what is important and what is not, what seems to be good strategy and what is poor. But we have to remember that, since chance plays a role in the outcome, the very same simulation, without changing any of the factors, will produce different results if tried a second time. Therefore, we have the difficult problem of estimating how much of success is due to good strategy and how much to pure luck.

There are certain rules of thumb. If the same model is run on the computer several times, under the same assumptions, the differences between the outcomes of several runs give us an estimate of the effect of chance. Decisions based on simulation models are rational only if one turns up differences due to changes in strategy which are significantly greater than those produced by chance. In the case of the baseball simulation, the differences due to changing line-ups turned out to be so small that only several hours of computation assured us that they were real.

Evaluation of simulation experiments may require very sophisticated statistical techniques. Therefore, we hope that a mathematical theory of simulation will be developed to make this powerful technique more widely applicable.

Markov Chains

A common feature of sciences in an advanced stage of development, notably of physics, is that theories are formulated in terms of concepts only indirectly connected with reality. For example, one cannot observe a gravitational field or entropy. Theories on this level of abstraction are rare in the social sciences, but the Estes-Burke learning model is of this type. The reader will find a detailed discussion in [1] and an elementary mathematical treatment in [5].

The theory is designed to predict behavior by men or animals in a very simple learning situation. The subject is presented with two possible choices of action on each trial, one of which is afterwards "rewarded" or designated as correct. The method of rewarding varies from experiment to experiment, and we are to predict the subject's long-range behavior as a function of what the experimenter does. The model has successfully explained the average reaction of a large group of subjects under varying schemes of rewards, even where the results are quite surprising.

We have to explain the model in some detail. Let us label the two possible choices as A and B. They could be "push the left button" versus "push the right button," or "turn left in the maze" versus "turn right," and so on. It is assumed that there is a set of *stimuli elements* that influence the subject's behavior. These must be thought of as theoretical constructs. At each stage, some of the stimuli elements are conditioned to A (inclined to an A decision) and the rest to B. When the subject has to make a choice, he selects a subset of the stimuli elements, according to a simple probabilistic scheme. If a of the elements selected are conditioned to A, and b to B, then there is probability $a/(a + b)$ that he will choose A. Thus, if we know the present status of the stimulus elements, we can predict the probability of an A response.

But how does the subject learn? If an A response is rewarded, all the stimulus elements he used for his decision become conditioned to A. Thus, the probability of an A response increases. If A turns out to be the wrong choice, that is, A is not rewarded, they are all conditioned to B, reducing the probability of an A response next time.

This is an example of a stochastic (probabilistic) model. It does not attempt to predict exactly what will happen on a given trial, but it assigns a probability to each possible outcome. This may be used to predict average behavior for a large number of subjects or to predict behavior in the long run. Clearly, we must make use of mathematical probability theory, but just how?

Let us assume that the experimenter has decided to reward an A response half the time and never to reward a B response. If we know the initial conditioning of the stimulus elements, we know the probability of each type of response. And we know the probability of a reward and, hence, of the selected stimulus elements being conditioned to A or B next time.

Everything of interest to the model is a consequence of how many of the stimulus elements are conditioned to A at a given time. And the above argument shows that, while we do not know for certain how this number will change in one trial, we know the probability for each possible new arrangement. Hence, we have a probabilistic process in which knowledge of the present state of the system completely determines the probabilities for the next time. Such a process is known as a "Markov chain."

The theory of Markov chains is barely sixty years old, and it has made tremendous progress within the last two decades. A subject pursued by many pure mathematicians, it is important also because of its wide variety of applications.

Let us consider what the pure mathematician can tell the learning theorist. First of all, we predict that there is an equilibrium for the conditioning of the stimulus elements and that this equilibrium will be approached. The equilibrium in question is a probabilistic equilibrium, quite different from equilibria experienced in mechanics. It is not that the stimulus elements tend to approach a given distribution — say $\frac{2}{3}$ conditioned to A and $\frac{1}{3}$ to B — but, rather, that the probabilities of the various possible distributions approach an equilibrium value. We may interpret this in terms of a large number of subjects: Given many subjects submitted to the same series of experiments, we can predict what percentage of them will have a given distribution after a large number of experiments. Hence, we can predict what percentage of them give an A response in the long run.

An alternative, and equally interesting, prediction is to forecast for a single subject how frequently he will make an A response in the long run. Let us illustrate this in terms of our example, where the experimenter rewards A half the time but never rewards B. If the subject were comletely rational and were given the opportunity to figure out a best strategy, he would always make an A response. But this is not what happens. He responds with A $\frac{2}{3}$ of the time, and with B $\frac{1}{3}$ of the time, precisely as we predict from the Markov chain model.

What the model predicts is that the subject unconsciously adjusts the frequency of his A response so that it will agree with the fraction of time that A is rewarded. In our example, let f be the fraction of time that he responds with A; hence, he responds with B $1 - f$ of the time. What does the experimenter do? Since an A response is rewarded half the time, but after a B response the experimenter always announces that A was right, the experimenter is rewarding A a fraction $(\frac{1}{2})\,f + (1 - f)$ of the time. This will agree with the subject's reaction only if

$$f = (\tfrac{1}{2})\,f + (1 - f),$$

which means that f must be $\frac{2}{3}$.

Other predictions of the theory, several of them equally surprising, are borne out by the experimental evidence.

A second interesting Markov chain model may be found in sociology. This model, due to B. P. Cohen, was designed to explain the results of some experiments by S. E. Asch (see [1] or [3]), which were made to test to what extent a subject can be pressured by a small group into conforming. The experimenter announces that he is going to test the subjects' ability to make certain visual comparisons. Actually, all but one member of the experimental group is in league with the experimenter. They agree on an incorrect answer, and then the lone subject is forced either to conform to or go against the unanimous opinion of the others.

Cohen's model is a simple Markov chain, in which the subject is assumed to be in one of four mental states before each trial:

1. Definite nonconformity.
2. Temporary nonconformity.
3. Temporary conformity.
4. Definite conformity.

It is assumed that he enters the experiments in state 2; that is, he is inclined to give the right answer but not permanently committed to this. He is assumed to move from state to state, according to a Markov chain process, until he reaches state 1 or state 4, which represent a permanent commitment. That is, he either rebels or decides to go along with the group.

We now have the framework of a model, but a number of parameters are left open. If the subject is in states 1 or 4, he stays there. But let us suppose that he is in state 2. What is the probability of moving to states 1 or 3 or of staying at 2? (It is assumed that he does not jump from 2 to 4 in one experiment.) This leaves 3 probabilities open whose sum must be 1 and, hence, we have 2 parameters to determine. The same holds for state 3, giving us a 4-parameter model.

Then Cohen proceeds just the way a physicist would: He estimates the parameters from the available data, using standard statistical methods, and shows that the resulting model gives a very good description of the behavior of a large number of subjects.

We have seen a number of simple examples of the applicability of mathematics to the social sciences. While most of the illustrations were necessarily oversimplified, they capture the flavor of the ways mathematics serves its newest customers. How often social sciences have to call on very recent mathematical discoveries! It leads us to believe that the best is yet to come.

BIBLIOGRAPHY

1. Berger, J., B. P. Cohen, J. L. Snell, and M. Zelditch, *Types of Formalization in Small-Group Research*, Houghton Mifflin, Boston, 1962.
2. D. Cartwright, F. Harary, and R. Z. Norman, *Structural Models: An Introduction to the Theory of Directed Graphs*, Wiley, New York, 1965.
3. J. G. Kemeny and J. L. Snell, *Mathematical Models in the Social Sciences*, Ginn, Boston, 1962.
4. J. G. Kemeny, *Random Essays*, Prentice-Hall, Englewood Cliffs, N.J., 1964.
5. J. G. Kemeny, J. L. Snell, and G. L. Thompson, *Introduction to Finite Mathematics* (second edition), Prentice-Hall, Englewood Cliffs, N.J. 1966.
6. A. Lotka, *Elements of Mathematical Biology*, Dover, New York, 1956.
7. Martin Shubik, *Strategy and Market Structure*, Wiley, New York, 1959.

BIOGRAPHICAL NOTE

John G. Kemeny is a professor of mathematics at Dartmouth College. He was born in Hungary and came to the United States as a boy before World War II. He was trained in pure mathematics, but when he came to Dartmouth in 1954, he was challenged by the Dean of the Faculty to develop a course specifically aimed at social-science students. One outcome was a course, designed with the help of social scientists, that has been widely copied in other schools. Another outcome, he says, was that "I and some others became convinced there were extremely interesting mathematical research problems coming out of the social sciences."

The structures of organic molecules are sometimes so bewilderingly complex that chemists have difficulty describing, classifying, and even naming them. Graph theory, a special tool borrowed from topology, has now been used to reduce even quite complicated chemical structures to a chain of numbers so that a computer can analyze them. This attempt to make organic chemistry more systematic could make it much easier for students to learn basic principles and to solve vexatious problems of classifying chemical compounds so that computers could be more readily applied to retrieve chemical information. It may be a forerunner of similar mathematical simplifications that will be applied to chemical genetics and other much more complex fields.

Topology of Molecules

Joshua Lederberg

The enterprise known as science rests on two pediments: the power and social utility of empirical knowledge, and the esthetic satisfaction that comes from an elegant restatement of principles. These views have been contrasted as the Baconian versus Newtonian justifications of science. Newton's name evokes a very apt image, his epochal contributions to the mathematical formulation of physics. Some esthetes judge how far a science has advanced in its development by the extent to which it has been mathematized — made into a deductive science by a set of axioms and rules for their manipulation.

The fruitfulness of pursuing such an aim is debatable for such fields as embryology, genetics, or psychology. Outside the rather special area of evolutionary theory, few examples of useful prediction are based upon any comprehensive mathematization of living behavior. On the other hand, for many special situations, models can be created that are sufficiently simplified to justify the application of some numerical mathematics or statistics. In his essay in this volume, Hirsh Cohen has discussed many examples of this kind of application of mathematics to biology and medicine.

With the rapidly growing speed, size, and availability of digital computers, the esthetic ideal of rationalizing a science acquires a new dimension of practical importance. If we could give biology sufficient formal structure, it might be possible to mechanize some of the processes of scientific thinking itself. Many of the most striking advances in modern biology have come about through the formulation of some spectacularly simple models of important processes, for example, virus growth, genetic replication, and protein synthesis. Could not the computer be of great assistance in the elaboration of novel and valid theories? We can dream of machines that would not only execute experiments in physical and chemical biology but also help design them, subject to the managerial control and ultimate wisdom of their human programmer.

This vision is so far beyond our present grasp, it makes what will be reported below seem quite trivial. These remarks may, however, give some notion of the reasons a geneticist took an interest in the formalization of organic chemistry. Chemical genetics embodies many statements in natural language, and its reasoning embodies an enormous range of expertise covering chemistry, geometry, and most of the natural sciences, as well as that most difficult realm, common sense. As a further complication, many quite fundamental discoveries are being reported almost daily. I wanted more experience with the mechanization of a simpler science before tackling chemical genetics. A scan across neighboring disciplines suggested that elementary organic chemistry might be a challenge that was more amenable yet had not been exhausted.

For various reasons, including the good fortune oi my association with Professor Carl Djerassi of Stanford's Chemistry Department, the analysis of mass spectral data for the solution of structural problems in organic chemistry was taken as the focal process for which a formalization would be attempted. Equally fortunately, Professor E. A. Feigenbaum joined the faculty of Stanford's Computer Science Department, and the entire effort of translating the formalisms and developing the heuristics for implementation on the computer has been done in close collaboration with him.

We may now turn to a consideration of the application of some elementary nonnumerical mathematics, that is, graph theory, for the representation of organic molecules. The use of these representations for a computer mechanization of the concepts of organic structural analysis will be summarized briefly.

The mathematical tool for translating chemical structures into a form that a computer can handle digitally is a concept that topologists call a *graph.* This kind of graph has little relation to the curves and bar charts used to display data; rather, it is a formal diagram for analyzing connections among a number of entities, in this case the individual atoms that make up an organic molecule.

Graphs have two components: *nodes* (representing atoms) and *edges* (chemical bonds between atoms). Each edge is associated with exactly two nodes, each node with at least one edge. The lengths of the edges are irrelevant. Disconnected graphs are regarded as representing molecules that are distinct, even if they are bound by diffuse chemical forces as in a crystal.

Our main approach is mapping: a rule of correspondence between a part of a chemical structure and a part of some abstract graph. Graphs lend themselves to canonical forms, that is, methodical choices among equivalent representations according to a precise rule, which eliminates ambiguity and redundancy. The objective is to represent each molecular structure by just one graph and, conversely, to have each graph represent just one structure. Chemistry will re-emerge after a few levels of abstraction.

The structural formula for an organic molecule is then a paragon of a topological graph, that is, the connectivity relations of a set of chemical atoms we take as the nodes of the graph. True, we recognize more than one type of connection — double, triple, and noncovalent bonds, as well as single bonds. From an electronic standpoint, however, the special bonds could just as well be denoted as special atoms. The structural graph does not specify the bond distances and bond angles of the molecule. In fact, these are known for only a small proportion of the enormous number of organic molecules whose structure is very well known from a topological standpoint.

Most of the syllabus of elementary organic chemistry thus comprises a survey of the topological possibilities for the distinct ways in which sets of atoms may be connected, subject to the rules of chemical valence. The student then also learns rules that prohibit some configurations as unstable or unrealizable. (He may later earn his scientific reputation by justifying or overturning one of these rules.) But the field of organic chemistry has reached its present stature without many benefits from any general analysis of molecular topology. These benefits might arise in applications at two extremes of sophistication: teaching chemical principles to college undergraduates and teaching them to electronic computers. They may also apply to the vexatious problems of nomenclature and systematic methods of information retrieval.

Although the topological character of chemical graphs was recognized by the first topologists, very little work has been done on the explicit classification of graphs having the greatest chemical interest. Some difficult problems, e.g., the analytical enumeration of polyhedra, remain unsolved.

This article will, then, review some elementary features of graphs that may be used for a systematic outline of organic chemistry.

All the Ways to Build a Molecule

A problem statement might be: Enumerate all the distinct structural isomers of a given elementary composition, say $C_3H_7NO_2$. That is to say, produce all the connected graphs that can be constructed from the atoms of the formula, linked to one another in all distinct ways, compatible with the valence established for each element (4, 3, 2, 1 for C, N, O, H, respectively). For compactness, H can be omitted from the representations, being implied by every unused valence of the other atoms.

The first discrimination is between trees and cyclic graphs, the "aliphatic" versus the "ring" structures of organic chemistry. Trees are graphs that can be separated into two parts by cutting any one link. How may we establish a canonical form for a tree, after first noting its order (number of nodes)?

The first step might be to find some unique place to begin the description. A tree must have at least two terminals and may have many more if highly branched; these are, therefore, not suitable starting points. However, each tree has a unique center. In fact, in 1869 Jordan showed that any tree has two kinds of center, a mass center and a radius center. Each center has a unique place in any tree; the two may or may not coincide.

To find the radius center, the tree is pruned one level at a time, cut back one link from every terminal at each level. This will leave, finally, an ultimate node or node pair (in effect, edge) as the center, the radius not of a length but, rather, of levels of pruning needed to reach the center.

To identify the mass center of a tree, we must consider the two or more branches that join to each nonterminal node. The center is the node whose branches have the most evenly balanced allocation of the remaining mass (node count) of the tree. This is the same as saying that none of the pendant branches exceeds half of the total mass. If the structure is a union of equal halves, the center is the edge that joins them.

Each of the centers (Figure 1) is unique and so could solve our problem of defining a canonical starting point of a description. The center of mass is more pertinent to finding a list of isomers, which of course have the same mass. The radius center is ill-adapted for this but matches conventional nomenclature, which is based on finding the longest linear path, that is, a diameter.

In chemical terms, the center divides the graph into two or more radicals. These radicals can be ordered by obvious compositional principles, giving rise to a canonical description of the whole graph in a linear code. Thus, methionine becomes (C(N)(C(O)(=O))(C—C—S—C)) or, in a parenthesis-free notation the example should make obvious, C· · ·NC· :OOC·C·S·C. This is more legible to the human reader, if the implied hydrogens are restored, as

CH· · ·NH2 C· :OH O CH2·CH2·S·CH3.

Any linear code has an implicit numbering system: Each atom is numbered according to the place where it occurs in the string.

Some thirty years ago, Henze and Blair showed how Jordan's principle could be used for the enumeration of isomers of saturated hydrocarbons and some simple derivatives of them. Here, the nodes are all carbon atoms, and the enumeration can proceed by working outward from smaller to larger complexes. For example, for the isomers of undecane, $C_{11}H_{24}$,

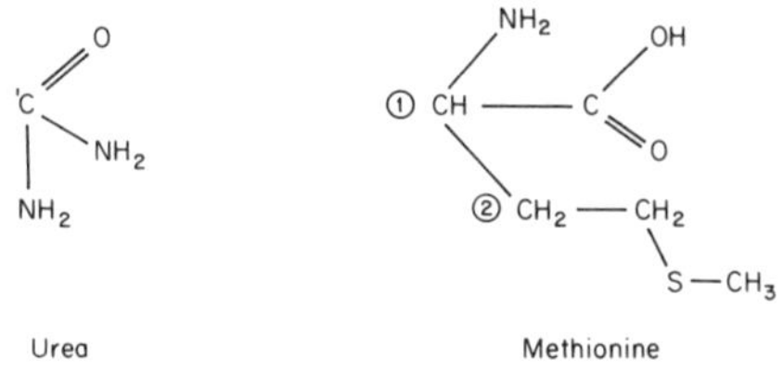

Urea Methionine

FIGURE 1. *Chemical trees and their centers. In urea, the carbon atom is both the radius center and the mass center.*

In methionine, carbon atom 1 is the mass center, according to the numerical partition 1 . . . 1 3 4. Carbon atom 2 is the radial center, on a diameter of 7, that is, the center of a largest string

(C—S—C—C—C—C=0).

For both analyses, we ignore hydrogen atoms.

one atom is designated as center, leaving 10 to be allocated among 2, 3, or 4 branches. Only the following partitions shown in Figure 1 satisfy the rules (leaving dissymmetry out of account):

Branches	Partitions	No. of partitions
2	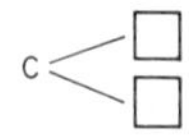 5 5	1
3 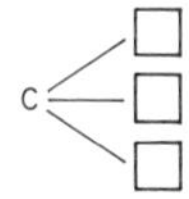	1 4 5; 2 3 5; 2 4 4; 3 3 4	4
4 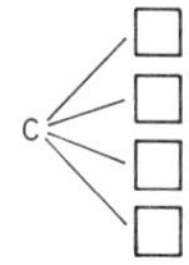	1 1 3 5; 1 2 2 5; 1 1 4 4; 1 2 3 4; 2 2 2 4; 2 2 3 3	6

No closed algebraic expression has been found for this enumeration. However, the recursive expansion was done manually by Henze and Blair with a few trivial errors later found by a computer check. No organic chemist will be surprised by the enormous scope of his field of study. There are, for instance, 366,319 isomeric icosanes, $C_{20}H_{42}$, and 5,622,109 icosanols, $C_{20}H_{41}OH$ (Table 1).

Table 1. Counting the different arrangements of compounds of carbon and hydrogen containing no double or triple bonds and no rings. These have the general formula C_nH_{2n+2}.

Number of Carbon Atoms		Number of Possible Isomers	
		11	159
1	1	12	355
2	1	13	802
3	1	14	1858
4	2	15	4347
5	3	16	10359
6	5	17	24894
7	9	18	60523
8	18	19	148284
9	35	20	366319
10	75		

The total range of acyclic compounds containing atoms other than that of the hydrocarbons C or H is, of course, very much larger than these subsets. To generate them, an allocation of nodes to constituent radicals takes account of the kind as well as number of remaining atoms. A complete enumeration of structural isomers of a given composition, for example of alanine, $C_3H_7NO_2$, can thus be made. We find 216 such isomers if we apply only these simple topological principles, compared with just 5 isomers of C_6H_{14}.

Graphs of Ring Compounds

Cyclic graphs are less tractable than trees. A linear representation is difficult because every path may return to a specific node already defined. The symmetries of cyclic graphs complicate the problem of defining a unique center on morphological criteria. These taxonomic difficulties are reflected by the existence and popularity of the American Chemical Society's Ring Index, which displays the "11524 rings known to chemistry" together with a profusion of synonyms and arbitrary numbering systems. Many more rings are discovered every day.

Molecules may also contain both acyclic and cyclic parts. However, if a strictly cyclic part is once defined, it can be regarded as a single node in a tree.

We now consider the strictly cyclic graphs, wherein at least two (sometimes more) links must be cut to separate the graph. First we produce a set of strictly trivalent cyclic graphs. Then these are related to the chemical graphs by ignoring the bivalent nodes of the latter. That is, the trivalent vertices are preserved to describe an abstract, basic graph and each linear path between vertices maps onto an edge of the basic graph. The degenerate case of zero vertices, the circle, must be included in the set since the simple ring is the most important cyclic structure of organic chemistry. A double ring can be generated in only one way, mapping onto a two-vertex trivalent graph: the molecule naphthalene maps onto the hosohedron. Figure 2 gives some of the more familiar cyclic hydrocarbons to illustrate these correspondences.

Some organic molecules have one or more quadrivalent vertices. This contingency can be met head-on by enumerating the full set of corresponding tri- quadrivalent graphs. It is expedient to convert these, when needed, to trivalent graphs by any of a number of tricks. For example, map a quadrivalent node onto a pair of connected trivalent nodes.

We now proceed to enumerate the trivalent graphs, each with an associated canonical representation and an implied numbering of nodes and edges for mapping the molecule.

Once Around the Network

A practical key to the solution of this problem, as to many other network problems, takes advantage of the Hamilton circuits found in most of the abstract cyclic graphs having chemical interest. A Hamilton circuit is a round trip through the graph that traverses each node just once. It therefore uses n edges, leaving out $n/2$ edges of a trivalent graph. Figure 3 is Hamilton's own example, the dodecahedron, proposed by him as a parlor game, each node representing a city that the round-the-world traveler would wish to visit once but not more often.

A convenient representation of an HC maps the nodes and edges of the circuit as vertices and bounding edges of a regular polygon. The remaining $n/2$ edges then form chords, each node being one of the two termini of one chord. A description of the graph then needs only some notation for the $n/2$ chords. First, we should canonicate the orientation of the polygon, having chosen to initialize the HC arbitrarily among n nodes and 2 directions (the rotational and reflectional symmetries of the polygon). Each node is joined by some chord having a certain span. The span list can be put in cyclic order, where it is immaterial which node is selected as starting point. The effect of reflection is also easily computed. If the span list is regarded as a number, its minimum value under rotation or reflection becomes the canonical form. For example, an 8-node graph might be represented (Figure 4) by any one of the span lists 17522663, 31752266, and so on, or the reflections 75226631, and so on. Of these, one quickly finds that 17522663 is the lowest-valued, hence the canonical form.

The same procedure establishes a canonical ordering of the nodes and edges. For the latter, we take the HC sequence (the polygon) first, then each chord in order of first reference.

The span list has n terms. Only $n/2$ are necessary since each chord is referred to twice in the span list. For an abbreviated code, simply omit the

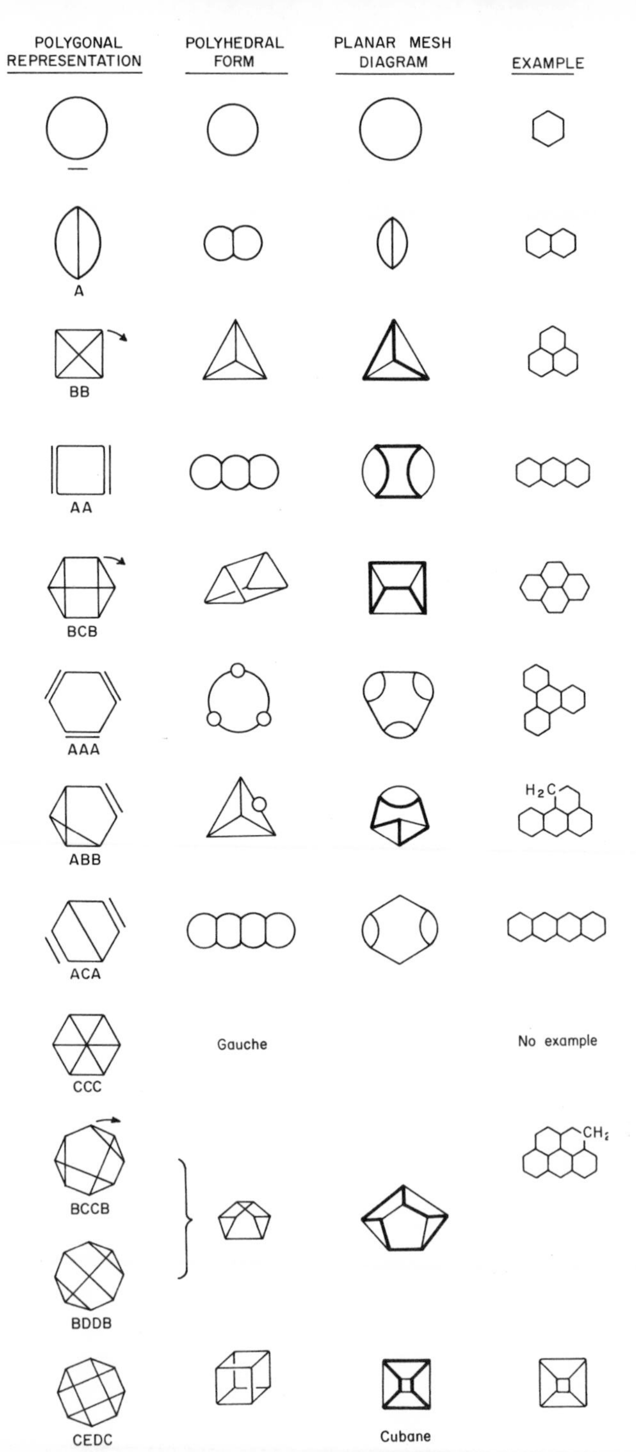

FIGURE 2. *The cyclic trivalent graphs with 8 or fewer nodes. Up to 6 nodes, these all have Hamilton circuits but may also be represented in other ways. In a few examples, the circuits are drawn with emphasis on planar map representations. Complete tables of chord lists like those shown under the circuit (polygonal) representations have been published for up to 12 nodes, virtually exhausting graphs of chemical interest.*

The chemical examples are, wherever possible, hexacyclic hydrocarbons. Each vertex stands for a carbon atom.

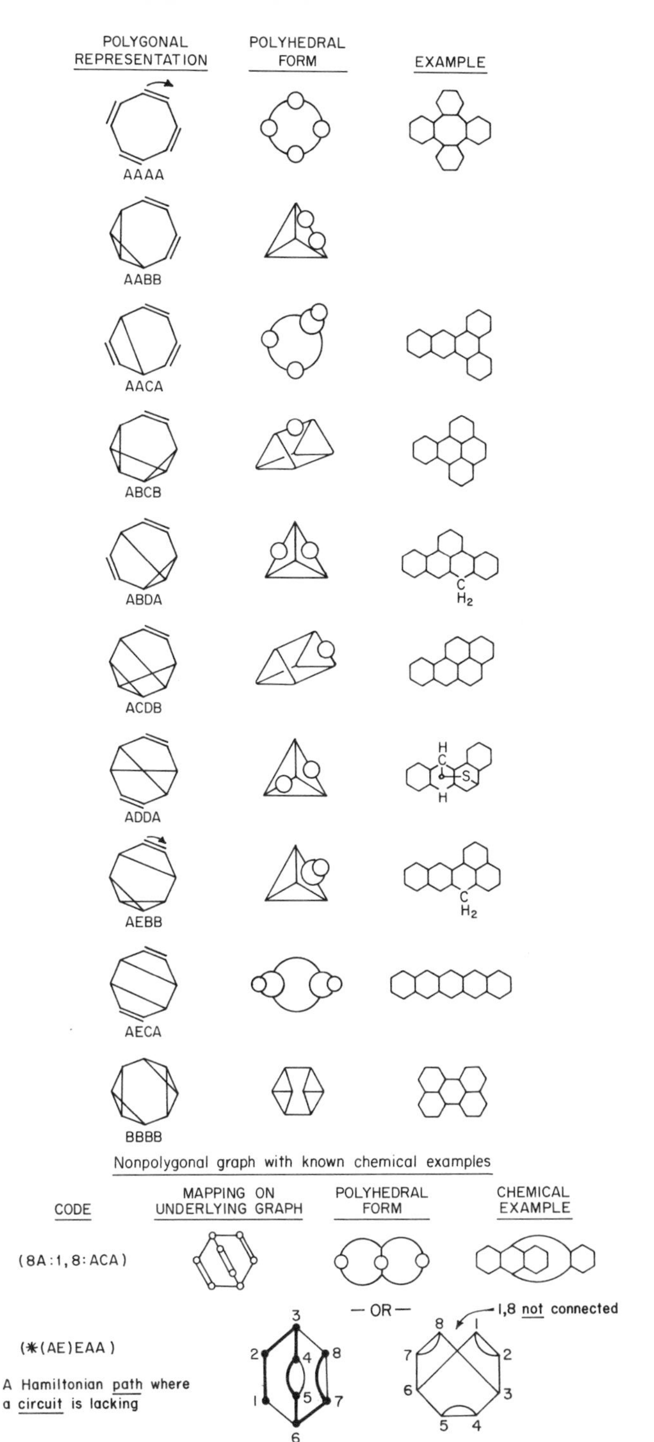

The final example has no Hamilton circuit. It can be computed either as a predicted union of two circuits (A with ACA, edge 1 with edge 8), in canonical form, or as a Hamiltonian path ((AE)EAA), the asterisk signifying that the polygon cannot be closed, and (AE) that two chords, A and E, both issue from the same, initial, node.*

As explained in the text, each chord of the polygonal representation is coded by one character for its span the first time it is encountered in a serial circuit of nodes.

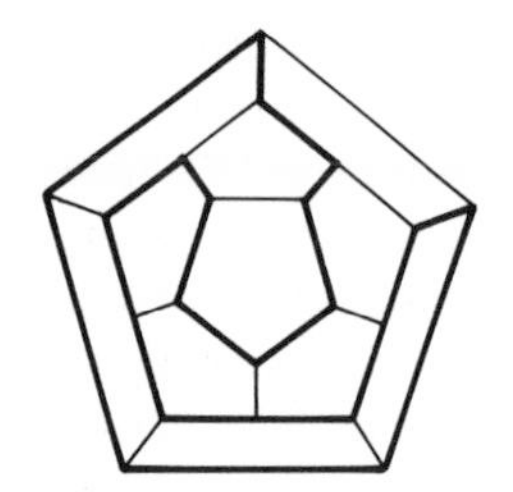

FIGURE 3. *Hamilton's own Hamilton circuit. The abstract dodecahedron, represented as a planar map of 20 nodes.*

second reference to any chord. Thus 17522663 becomes 1522 to encode the graph in a canonical form (Figure 4). Since we need more than 10 numbers, we use the alphabet, character by character. Thus 1522 becomes

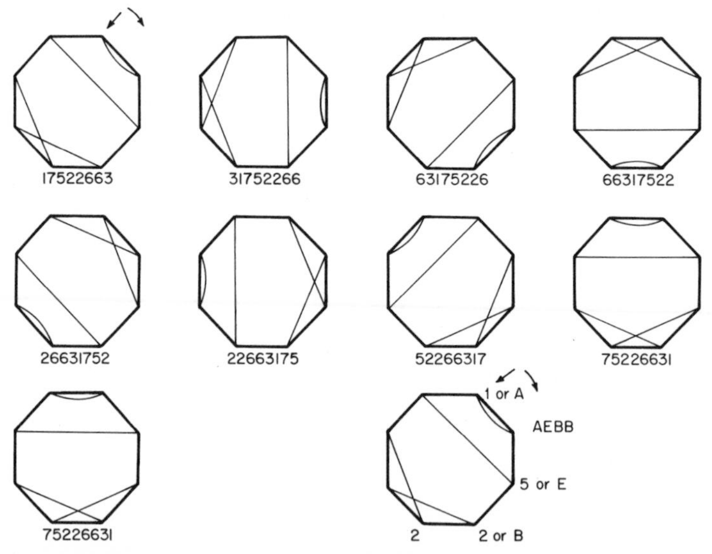

FIGURE 4. *Symmetries and encoding of a cyclic trivalent graph with 8 nodes. There are 16 symmetry operations (8 rotational × 2 reflection). Shown are 8 rotations, and a reflection that could be combined with each of these. With each figure is also a span list; the canonical choice of the 16 (not all distinct) is the lowest-valued span list, 17522663, calculated with the upper rightmost node as the initial. This can then be reduced to the code AEBB.*

AEBB. Furthermore, we can reconstruct the graph from the code by retracing the steps just recited. *Caution*: Unlike span lists, the abbreviated chord lists cannot be freely rotated.

Having a systematic, linear code, we are now in a position to compute all possible Hamilton circuits. Any span list is a string of numbers; therefore, the complete set of circuits can be sieved by a computer program from the series of integers. A great deal of fruitless computation can be saved by incorporating some of the canons of preferred representations into the generating algorithm. For example, no later digit can be smaller than the leading digit; else a simple rotation of the span list, which is an obvious isomorphism, would give a smaller, preferred code number.

In this manner, exhaustive lists of Hamilton circuits for $n \leq 12$ have been computed. They are illustrated here up to $n = 8$ (Figure 2). Some planar, trivalent graphs lack a Hamilton circuit. The simplest has 8 nodes (Figure 2, last item) and, as it happens, it does underlie the mapping of a known compound. Obviously, these graphs will not be anticipated by a computer program that generates Hamilton circuits. However, it is not difficult to describe these figures as unions of circuits or else, for every practical case, as Hamilton paths. Furthermore, at each level of graph-building, it is possible to anticipate combinations of cut edges that will yield circuit-free graphs upon union with other partial graphs. A complete set of trivalent graphs is, therefore, computable.

The special case of the smallest, circuit-free trivalent polyhedron has been a challenge to mathematicians for some time. A polyhedron is here defined as a 3-connected trivalent planar graph, that is, one that cannot be separated with less than three cuts. Tait had conjectured that a Hamilton circuit always existed, but this was refuted by Tutte with a 46-node counterexample. Subsequently a 38-node case was built which lacks a Hamilton circuit (Figure 5). So far as is known, this is the smallest;

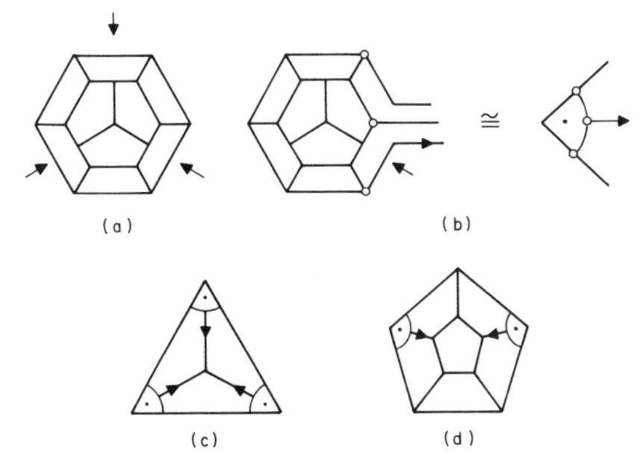

FIGURE 5. *A graph with special edges and two HC-free polyhedra. (a) has 16 nodes. The marked edges are included in any HC of the graph. Hence the 3-cut (b), with 15 nodes, obligates the marked edge as part of an HC of any graph in which (b) is inserted. This leads to a contradiction, that is, no Hamilton circuit in (c) Tutte's graph, with 46 nodes and (d) with 38 nodes.*

however, there is no proof of it. All the trivalent polyhedra with up to 18 and 20 nodes have been scrutinized or anticipated, and all have Hamilton circuits.

No incisive theory yet deals with these curiosities of empirical mathematics, in the same sense that we have no systematic generator for producing the *n*th prime number. However, if the elegance of the theory of polyhedra is marred by such empiricism, it is no impediment to putting the chemistry of real molecules on the computer.

Nonplanar graphs are theoretically important possibilities. The corresponding molecules (Figure 6) should be difficult but not impossible to synthesize. So far, none has been reported.

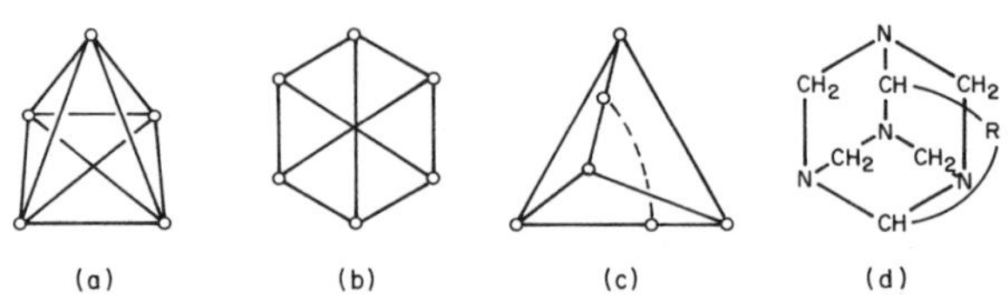

FIGURE 6. *Nonplanar graphs. (a) and (b) are Kuratowski's fundamental forms, 4-valent and 3-valent, respectively. At least one of these must be included in any nonplanar graph. (c) is a projection of (b) as a tetrahedron with an additional internal chord, and (d) is a hypothetical molecular structure that maps on to (c).*

Mapping and Symmetry

Having explored the trivalent graphs, we now return to mapping chemical atoms on their nodes and bonds or linear chains on their edges. Many graphs have substantial symmetry, and the correspondingly redundant operations must be considered to decide on a canonical representation. Here, again, the HC's are helpful. If an HC is present, it can also be projected on the same graph after any symmetry operation. Therefore, the whole set of symmetry operations is included within the list of the HC's, giving both remarkable economy of computational effort to the search for the symmetries and a straightforward expression of the operators. To describe a molecular structure, we can map it on an arbitrary choice of form and then subject the result to the symmetry operators. The canonical representation satisfies some rule, say the highest-order listing, of the mapped elements. Thus, for the morphine nucleus, we would have to choose among the 4 symmetries of its underlying graph (Figure 7), and we can then encode the morphinan molecule as

(8BDDB 4*0031301000 NC3,C3,O,C3,C).

The first two words define the basic map, "*" standing for a fused edge, and the digits for the lengths of the paths between vertices. The last clause maps the atomic strings on to the nonempty edges.

Besides the linear paths of the cyclic structure, the mapping may also include specifications for fused edges (quadrivalent centers), heteroatom replacements of vertices, and specifications of stereoasymmetry of vertices. The details are inevitably fussy, but the computer handles all the fuss once the program is worked out. After the mapping, each atom is numbered in the order of its reference.

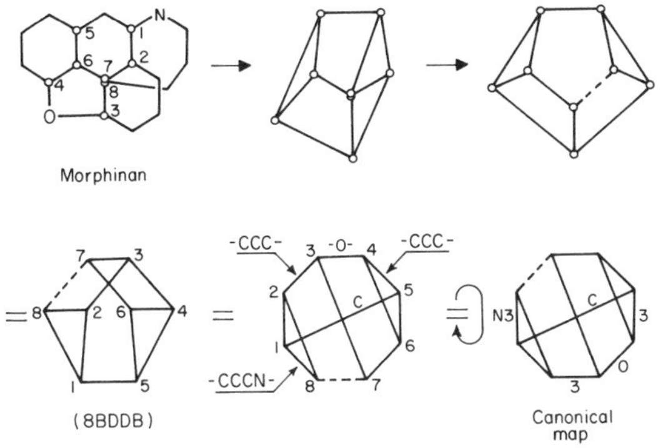

FIGURE 7. *Mapping a complex ring: morphinan.*

Applications

This development was needed for a continuing effort to program the automatic computation of structural hypotheses to be matched against various sets of analytical data, especially mass spectra. The growing sophistication of instrumental methods has already begun to outdo the chemist's capacity to interpret the results. Since mass spectrometers now commercially available can generate 10,000 spectra per second, the need for computational assistance to make full use of this speed is self-evident. Such devices are also being considered for the automated exploration of the planets, which puts even heavier demands on the local intelligence available to the system.

These applications relate primarily to the possibility of anticipating hypothetical structures. The language also provides a format for expressing synthetic insights, that is, the elementary reactions by which functional groups can be altered or exchanged. We might then expect the ultimate development of computer programs that have been taught a few thousand unit processes (and their limitations) and could be challenged to anticipate a synthetic route from given precursors to a given end product. Such programs might at least assist the chemist by reminding him of a few among myriad possibilities of combining the unit processes learned from the same chemist or, better, from a diverse school. For the moment, we do not consider the empirical testing in the computer's own laboratory of a few thousand routes chosen on its own initiative.

The nomenclatural utility of a system of canonical forms is self-evident. We are very nearly at the point where linear notation may again be dispensable for human use since the computer should be able to interpret structural graphs as such. However, a mathematically complete system of classification of structures is still important, regardless of the notation in which the structures are expressed.

There are, of course, many alternative approaches to notation, reviewed by a National Academy of Sciences Committee (1964) and appearing from time to time in the *Journal of Chemical Documentation*. As far as I know, none of them has been addressed to the exhaustive prediction of canonical forms, and most of them are too complicated to be easily adaptable to this end.

Computer Implementation

The notation of the computer language called DENDRAL is the foundation of some current efforts at mechanized induction in organic chemistry. A program to generate all isomers of tree structures has been fully implemented in the LISP programming language and is routinely run on a time-shared PDP-6 computer at Stanford University. Most of this program was developed on the Q-32 computer of the System Development Corporation at Santa Monica, California, using remote teletype consoles located at various homes and offices at Stanford, 400 miles away.

The kernel of the program is a "topologist" embodying the principles of the first part of this paper. It is, however, restrained by some common-sense chemistry to eliminate many inappropriate constructions. For example, the chemist knows that enolic structures like ·CH=CH·OH are unstable, rapidly reverting to a tautomeric equivalent (aldehydes), ·CH2·CH=0, and this information is embodied in the higher-level program. Also included is a model of the process of molecular fragmentation in the mass spectrometer, leading to a deduction of the mass spectrum expected from a hypothetical structure. The program uses the input data to guide its induction of candidate hypotheses, then tests these hypotheses deductively against the data, in an emulation of the traditional scientific method.

Much to our surprise, the program already works with real data, sometimes giving correct solutions. Not so surprising, the program greatly outdoes human chemists in problems like generating all the isomers of a given composition. Most of us founder on the isomorphisms.

Students encountering organic chemistry for the first time are often frustrated because they are challenged with graph-theoretic concepts, implicitly, without being told that this is their problem. For example, a student is expected to use his intuition to discover that there are only two isomers of C_2H_6O (in our notation, CH2··CH3 OH, ethanol, and O··CH3 CH3, dimethyl ether), but this intuition is achievable only with extensive practice. And even an experienced chemist will be hard-put to describe, irredundantly, all the isomers of slightly more complicated molecules, say $C_6H_{14}O$. Many problems in elementary chemistry are solved by excluding all but one of a list of possible isomers, implying that the whole list is deducible. The concept of the center of a tree and the algorithms for systematic generation of isomers should be of substantial value in teaching this subject, quite apart from the implementation of the algorithms on the computer. The same consideration should also apply to the ways in which rings can be built and to positional isomerism of substituted rings.

BIBLIOGRAPHY

E. F. Beckenbach, *Applied Combinatorial Mathematics*, Wiley, New York, 1964.
O. Ore, *Graphs and Their Uses*, New Mathematics Library, Random House, New York, 1963.
J. Lederberg, "Topological Mapping of Organic Molecules," *Proc. Nat. Acad. Sci. U.S.*, *53*, 134–139, 1965.
J. Lederberg and E. A. Feigenbaum, "Mechanization of Inductive Inference in Organic Chemistry," in *Formal Representation of Human Judgment*, B. Kleinmetz, Ed., Wiley, New York, 1968.
V. Klee, "Long Paths and Circuits on Polytopes," in *Convex Polytopes*, B. Grunbaum, Wiley, New York, 1967.
J. Lederberg, "Hamilton Circuits of Convex Trivalent Polyhedra (Up to 18 Vertices)," *Amer. Math. Monthly*, *74*, 522–527, 1967.

BIOGRAPHICAL NOTE

Joshua Lederberg is Professor of Genetics at Stanford University School of Medicine, Palo Alto, California. He interrupted his medical studies at Columbia University in 1946 for what was intended to be a brief research leave with Professor E. L. Tatum at Yale University. This work on the genetics of bacteria became too fruitful to be dropped, and indeed was the basis of the Nobel Prize in Medicine that was awarded to him twelve years later.

Professor Lederberg's best-known research accomplishment has been the discovery of mechanisms of genetic exchange in bacteria, which underlie much contemporary research in molecular biology and which have helped to unify modern concepts of life on earth. He has been involved in the search for extraterrestrial life that has become part of the space-exploration program. As indicated by this essay, he is also interested in scientific methodology as a problem in itself, with a view to augmenting human intelligence by the use of machines.

He has been broadly concerned about the implications of new biological knowledge for mankind — for example, in his role since 1961 as Director of the Lt. Joseph P. Kennedy, Jr. Laboratories for Molecular Medicine (dedicated to the study of mental retardation). He has served on President Kennedy's Panel on Mental Retardation, and is currently a member of the National Advisory Council of the National Institute of Mental Health. He also writes a column on "Science and Man," which appears weekly in the *Washington Post* and is syndicated to a few other newspapers on four continents.

Euclid himself was uneasy about one of the pillars of his geometry, the familiar postulate about parallel lines. After trying without success to prove this supposedly self-evident truth, later mathematicians finally began to wonder what would happen if they discarded it in favor of other assumptions. The result of their experiments was non-Euclidean geometry, which is in some respects startlingly different from the Euclidean kind but no less logical and consistent. The success of non-Euclidean geometry stimulated mathematicians to use their imagination much more freely. As a consequence, geometry today is concerned with many kinds of spaces, some of which defy intuition. Though non-Euclidean geometry may seem peculiar at first glance, it may actually turn out to be the best way to describe the cosmos.

Non-Euclidean Geometry

H. S. M. Coxeter

Euclidean geometry, the kind traditionally studied in high school, goes back to what has been called the most influential and widely used textbook of all time: the celebrated *Elements* of the Greek mathematician Euclid. Written around 300 B.C., the *Elements* began by distinguishing a few basic notions (point, line, and so on) and listing in the form of ten postulates certain properties of these notions that were to be accepted at the outset. Then came a systematic development of assertions (propositions or theorems) that could be demonstrated to follow logically from the postulates. These postulates were viewed as expressing "self-evident truths" about the concept of space, dictated by man's experience of the physical world. As we shall see, this view persisted into the nineteenth century.

Nevertheless, even Euclid himself seems to have looked upon some of these postulates as "more self-evident" than others. He was especially reluctant to employ his Fifth Postulate, and in fact he made no appeal to this postulate in proving his first twenty-eight propositions. The Fifth Postulate says that, if (in a plane) two points R and M are on the same side of a line PN and if the angles $\angle MNP$ and $\angle NPR$ total less than 180°, the rays PR and NM, extended sufficiently far beyond R and M, will meet (see Figure 1). The reason for uneasiness about the Fifth Postulate as a

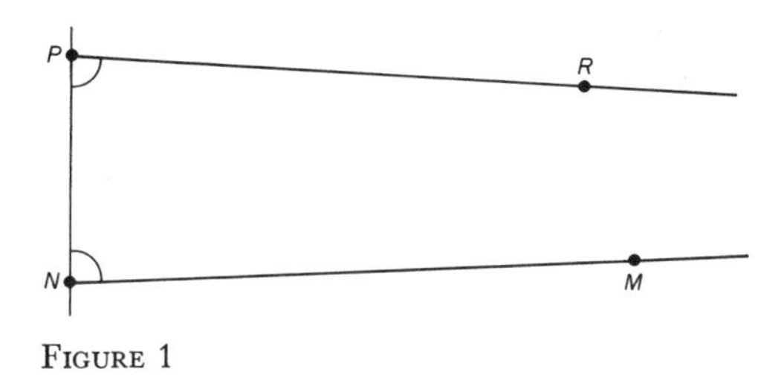

FIGURE 1

valid idealization of physical experience is tolerably clear. In some cases, the meeting point of these rays must be thought of as "very far out in space," going beyond the realm of direct physical experience.

Between Euclid's time and the nineteenth century, mathematicians sought a more intuitively convincing but equivalent postulate, one which together with Euclid's other postulates would yield the same body of theorems. Several equivalent postulates were found, including one, due to John Playfair (British, 1748–1819), now commonly used in high school courses and called the *parallel postulate.* It states that, given a line l and a point P not on l, the plane of l and P contains only one line passing through P and not meeting l (that is, parallel to l). This was only seemingly more satisfactory, for actually, of course, the very notion of parallel lines, which never meet "no matter how far extended," already involves the conception of what happens very far out in space.

Hyperbolic Geometry

During the long period from the time of Euclid to the nineteenth century, many professional and amateur mathematicians also tried to deduce the disquieting Fifth Postulate from the others by the indirect method: They hoped that its denial would lead to a contradiction. The most persistent of these pioneers was the Jesuit Gerolamo Saccheri (Italian, 1667–1773) who discovered, for the sole purpose of demolishing them, many theorems of what we now call *hyperbolic geometry.* It was C. F. Gauss (German, 1777–1855), generally considered one of the three or four greatest mathematicians of all time, who first took the modern standpoint that no absurdity or contradiction could result from hyperbolic geometry, in which Euclid's Fifth Postulate is replaced by the following one: For any line l and any point P not on it there is an angle $\angle OPQ$ such that the only rays from P that meet l lie inside this angle (Figure 2). The bounding rays PQ and PO are said to be *parallel* to l. It turns out that there are then infinitely many lines through P (such as u and all others in the supplement of $\angle OPQ$) which, like the parallels, fail to meet l; these are said to be *ultraparallel* to l. Hyperbolic geometry is one of the two geometries that we call *non-Euclidean.* The other, *elliptic geometry*, will be described later on.

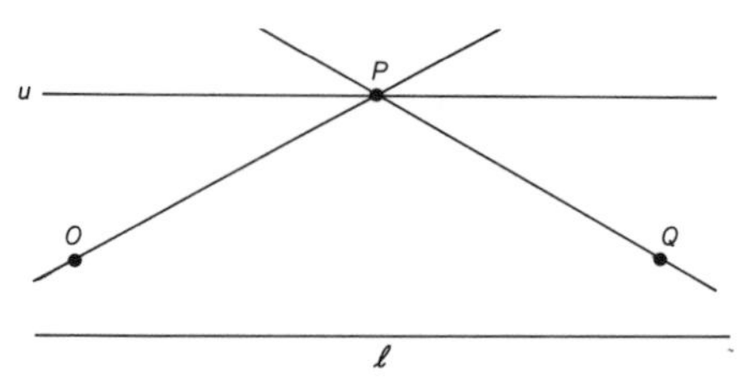

FIGURE 2

In hyperbolic geometry every line segment PN of length a (see Figure 3) determines an acute *angle of parallelism* $\Pi(a)$ such that, if $\angle MNP = 90°$ and $\angle NPR \geq \Pi(a)$, the rays PR and NM will not meet even though $\angle NPR$ may be acute. Moreover, in this geometry, the angles of a triangle ABC satisfy the inequality $A + B + C < 180°$, and the area of the triangle is

$$\left(1 - \frac{A + B + C}{180°}\right)\pi.$$

The effect of the discovery of hyperbolic geometry on our ideas of truth and reality has been so profound that we can hardly imagine how shock-

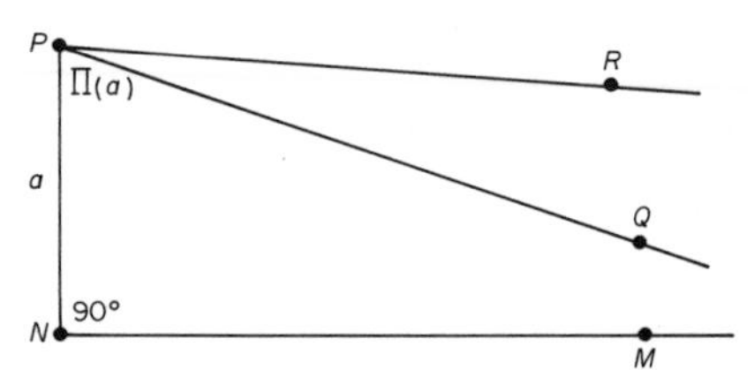

FIGURE 3

ing the possibility of a geometry different from Euclid's must have seemed in 1820. Gauss wrote about this "anti-Euclidean" geometry in letters to his friends but apparently preferred to avoid the controversy that might have resulted from publishing it. Between 1823 and 1826, it was rediscovered by John Bolyai (Bolyai János, Hungarian, 1802–1860) and N. I. Lobachevsky (Russian, 1793–1856), who share the credit for the first published accounts. They developed the subject into an impressive collection of theorems and formulas, such as

$$\tan \tfrac{1}{2}\Pi(a) = e^{-a}.$$

Lobachevsky became a professor at the University of Kazan but was never made an Academician. Bolyai, though fully aware of the significance of his discovery, received even less recognition during his lifetime. In 1848 he read a German version of Lobachevsky's work and was full of admiration. However, still smarting from unkind treatment he had long ago endured from Gauss, he never ventured to write to the prosperous Russian. There is no evidence that Lobachevsky was aware of Bolyai's existence.

In some places, the widespread attitude of intolerance lingered on for many years. C. L. Dodgson (English, 1832–1898), alias Lewis Carroll, repudiated hyperbolic geometry in 1888 as being too fanciful. It is a strange paradox that he, whose Alice in Wonderland could alter her size by eating a little cake, was unable to accept the possibility that the area of a triangle could remain finite when its sides tend to infinity.

Elliptic Geometry

Hyperbolic geometry resulted from denial of Euclid's Fifth Postulate. *Elliptic geometry* results from denial of his Second Postulate. The Second Postulate says that a straight line can be extended indefinitely; this statement clearly means that a line is infinite and not closed. But in elliptic geometry every line is finite though unbounded; that is to say, it has a finite length and is closed like a circle. On the surface of a sphere (such as the geographical globe or the earth itself), the shortest distance between two points is an arc of a great circle. It was natural for the navigators of ancient times to regard such arcs as the "line segments" of a special kind of two-dimensional geometry, namely *spherical geometry:* the geometry of figures drawn on the surface of a sphere of unit radius. We shall see later that this is almost the same as elliptic geometry. It was studied by Menelaus of Alexandria about A.D. 100 and by the Arabs about A.D. 1000. Its most famous theorem states that the three angles of a spherical triangle ABC satisfy the inequality $A + B + C > 180°$ and that the area of the triangle is

$$\left(\frac{A + B + C}{180°} - 1\right)\pi$$

(Albert Girard, 1595–1632).

The round shape of the earth had been recognized by unprejudiced people as early as about 240 B.C., when Eratosthenes computed the diameter with remarkable accuracy. The surface of the earth has a finite area although we can travel on it straight ahead forever. In other words, the domain of spherical geometry is finite but unbounded. Between 1852 and 1854, Ludwig Schläfli (Swiss, 1814–1895) and Bernhard Riemann (German, 1826–1866) independently conceived the possibility of extending this idea to three (or more) dimensions. Just as the line can be finite though unbounded, the unboundedness of space does not necessarily imply infinite volume. A sufficiently powerful telescope could conceivably enable an astronomer to see the back of his own head, except for the slight complication that the light reflected from his head would take thousands of millions of years to reach his eye. This idea, that space could be unbounded without being infinite, was fruitfully utilized by Albert Einstein (German, 1879–1955) in the creation of his general theory of relativity.

Riemann was a superb mathematician whose profoundly original ideas deeply affected the course of both geometry and analysis and foreshadowed the development of topology. The name "Riemannian geometry" is used for a highly sophisticated subject of which spherical geometry is only a very simple special case. Schläfli, too, made other contributions to geometry and analysis. One of his later achievements was a formula for the volume of a spherical tetrahedron, which is enormously harder to find than Girard's formula for the area of a spherical triangle.

It was Felix Klein (German, 1849–1925) who first saw how to remedy the awkward property of spherical geometry that two lines in a plane (being two great circles of a sphere) have not just one but two common points. Since every point determines a unique antipodal point, and every figure is thus duplicated at the antipodes, he realized that nothing would be lost but much gained by abstractly identifying each pair of antipodal points, that is, by changing the meaning of the word "point" so as to call such a pair one point. In other words, the points of the so-called *elliptic plane* are represented by the pairs of antipodal points on the unit sphere. It follows that all the properties of projective geometry belong also to elliptic geometry. So do all the "local" properties of spherical geometry; for example, the area of a triangle ABC is still

$$\left(\frac{A+B+C}{180^\circ}-1\right)\pi.$$

On the other hand, since the surface of the unit sphere amounts to 4π, the area of the whole elliptic plane is only 2π.

We began by defining elliptic geometry as the result of modifying Euclid's geometry by denying his Second Postulate. More precisely, the new version of this postulate is the statement that every two lines in a plane have just one common point. In other words, there are no parallels in such a plane. Despite this fact, W. K. Clifford (English, 1845–1879) found pairs of skew lines in elliptic space which are "parallel" in the sense that, if a variable point P runs along one of the lines while N is the

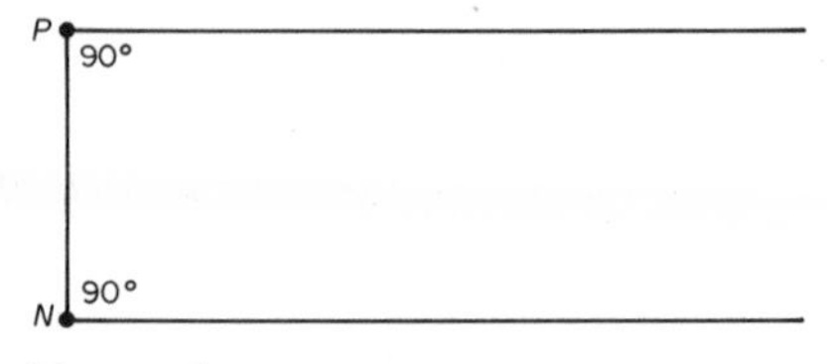

FIGURE 4

nearest point on the other (as in Figure 4), the length PN remains constant. The precise connection between projective geometry and elliptic geometry was first formulated by another Englishman, Arthur Cayley (1821–1895).

It is reasonable to regard Euclid's Second Postulate as including the familiar properties of order, the idea of one point lying between two others. Here, too, elliptic geometry behaves differently: Instead of *intermediacy*, it has *separation*, the idea that two pairs of points on a line may separate each other.

What Happens Far Away?

The use of the names "elliptic" and "hyperbolic" for the two kinds of non-Euclidean geometry is due to Klein, and may be explained by the following appeal to common sense. Consider the very simple arrangement (Figure 4) of two rays going out in the same direction from the two ends of a line segment PN perpendicular to both of them. The distance between the rays obviously remains equal to PN wherever we can measure it. But imagine the rays to be extended so far out into space that direct measurement is no longer possible. What right have we to assume that the distance between the rays may not eventually decrease or increase? If, in fact, it decreases, the geometry of space is elliptic (from the Greek *elleipein*, to fall short). If it increases, the geometry is hyperbolic (from *hyperballein*, to throw beyond).

Suppose the point S (Figure 5) represents a star, while N and P repre-

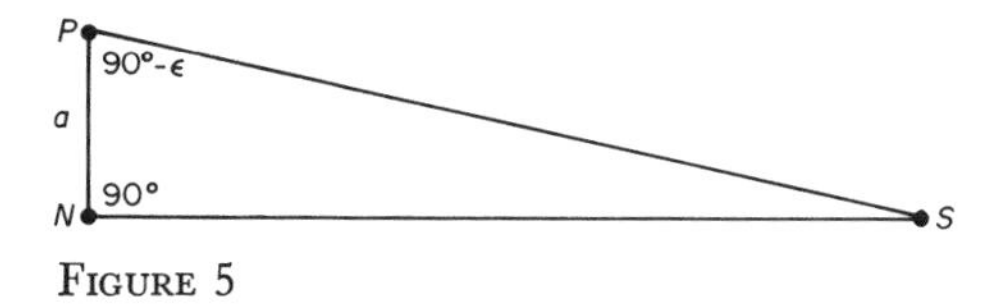

FIGURE 5

sent the positions of an observer at two times, six months apart, so that the distance $a = PN$ is the diameter of the earth's orbit. If $\angle SNP = 90°$, $\angle NPS$ falls short of $90°$ by a small angle called the *parallax* of the star. If astronomical space is Euclidean, the parallax

$$\epsilon = 90° - \angle NPS = \angle PSN$$

enables us to compute the distance of the star by the formula $NS = a \cot \epsilon$. If the space is elliptic, a sufficiently distant star should have a negative parallax; if hyperbolic, the parallax of every star should exceed a particular value, namely,

$$90° - \Pi(a).$$

The fact that every observed parallax is positive (or too small to be distinguished from zero) proves not that our physical universe is Euclidean but merely that, if it is elliptic, a star of negative parallax is too far away to be seen and that, if it is hyperbolic, the difference between $\Pi(a)$ and 90° is too small to be measured.

A further complication is that, in observing a distant star, we are receiving light that left it millions of years ago. Thus, space and time are inseparable. In fact, the geometry of the space-time continuum is so closely related to the non-Euclidean geometries that some knowledge of them is an essential prerequisite for a proper understanding of relativistic cosmology.

Logical Consistency

It is interesting to reflect that, although Gauss, Lobachevsky, and Bolyai lived and died with the intuitive certainty that their geometry was just as logically consistent as Euclid's, they possessed no rigorous proof of it. Such proofs, for elliptic as well as hyperbolic geometry, were supplied by Klein, Eugenio Beltrami (Italian, 1835–1900) and Henri Poincaré (French, 1854–1912). For instance, Beltrami discovered in 1868 that the lines of the hyperbolic plane can be represented in the Euclidean plane by the chords of a circle, with definitions for angle and distance suitably modified to give the effect that the points on the circumference are infinitely far away from any proper (that is, interior) point. Two lines are parallel if the representative chords meet on the circumference, ultraparallel if they meet outside. In this "model," Figure 2 becomes Figure 6,

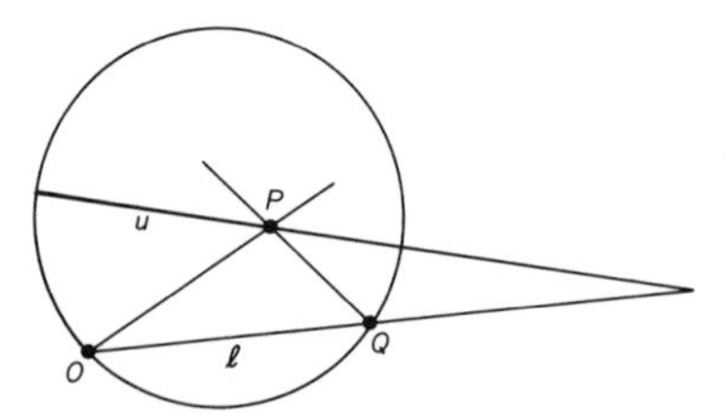

FIGURE 6

and a triangle inscribed in the circle represents a hyperbolic "triangle" whose infinitely long sides are parallel in pairs, so that the three angles are all zero and the area is π.

In our time, geometers are still exploring these new Wonderlands, partly for the sake of their applications to cosmology and other branches of science but much more for the sheer joy of passing through the looking glass into a land where the familiar lines, planes, triangles, circles, and spheres are seen to behave in strange but precisely determined ways.

BIBLIOGRAPHY

Roberto Bonola, *Non-Euclidean Geometry*, Dover, New York, 1955.

H. S. M. Coxeter, *Non-Euclidean Geometry*, University of Toronto Press, Toronto, Canada, 1965.

László Fejes Tóth, *Regular Figures*, Pergamon, Oxford; Macmillan, New York, 1964.

Stefan Kulczycki, *Non-Euclidean Geometry*, Pergamon, Oxford; Macmillan, New York, 1964.

Heinrich Liebmann, *Nichteuklidische Geometrie*, Göschen, Leipzig, Germany, 1923.

BIOGRAPHICAL NOTE

H. S. M. Coxeter is a professor of mathematics at the University of Toronto. A former president of the Canadian Mathematical Congress and vice president of the Mathematical Association of America, he is a Fellow of the Royal Society. Professor Coxeter first became interested in non-Euclidean geometry while browsing in his high school library. It appealed to him because it was enough like Euclidean geometry to be visualized, yet sufficiently different to be stimulating.

The problem of drawing useful inferences from observations is something like probability theory inside out. In probability problems, you know how nature works and want to find out what is likely to happen. With statistical inference, on the other hand, you observe what has happened and then try to deduce the mechanism that made it happen. Statistical inference is inevitably fallible. One of the statistician's main tasks is to lessen the chance of guessing wrong. This requires him to recognize subtleties that go beyond purely mathematical considerations but must somehow be incorporated into mathematical procedures. In order to decide which of several procedures will best serve the user, the statistician must abandon time-honored rules of thumb and rely on precise analysis of the problem. That, the author explains, is what the modern mathematical theory of statistical inference is all about.

Statistical Inference

J. Kiefer[1]

The word "statistics" conjures up many meanings. It suggests the state of the economy, Detroit's quality-control methods, medical evidence linking lung cancer with smoking, and the physicist's configurations in statistical mechanics. In this paper, our interest is limited to *statistical inference*. This is the study of various possible procedures for analyzing data in order to guess the nature of the physical or biological mechanism that produced these data. It is an applied area of mathematics with particular conceptual difficulties not found in other applied areas. The mathematical theory tries to give a rationale for selecting a procedure for analyzing the data rather than relying on intuition or rule of thumb. We cannot hope to develop any deep theoretical results of the subject here. We hope to treat only a few examples that may give you some feeling for subtleties in even the simplest statistical problems where practitioners tread — usually fearlessly, frequently by intuition, and quite often disastrously.

How does statistical inference differ from probability theory? In probability theory one specifies a model (chance mechanism) and studies its consequences. For example, suppose a coin has probability p of coming up "heads" on a single toss, and $1 - p$ of coming up "tails." Here p is a

[1] This article was prepared with support from ONR Contract No. NONR 401(50). Reproduction in whole or in part is permitted for any purposes of the U.S. Government.

number between 0 and 1 that reflects the physical makeup of the coin; roughly speaking, it is the approximate proportion of heads you would expect in a long series of tosses, and it would be $\frac{1}{2}$ for a "fair coin." Our model is to flip the coin five times, with independent flips. "Independence" means that successive flips have no probabilistic effect on each other, so that the probability of obtaining a particular sequence of outcomes on the five tosses is computed by multiplying the probabilities of the outcomes for the individual tosses. Thus, the probability of obtaining "heads, tails, heads, tails, tails," which we hereafter abbreviate HTHTT, is

$$p \times (1 - p) \times p \times (1 - p) \times (1 - p) = p^2(1 - p)^3. \tag{1}$$

Clearly, we obtain the same probability $p^2(1 - p)^3$ for any other sequence of five tosses containing exactly two H's, for example TTHHT. There are ten different arrangements of two H's and three T's, and the event "two heads in five tosses" occurs if the five flips come up in any of these ten arrangements, each of probability $p^2(1 - p)^3$. Thus, the event "two heads in five tosses" has probability $10p^2(1 - p)^3$. Similarly, the probability of obtaining exactly k heads in five tosses, hereafter denoted $b_p(k)$, is

$$b_p(k) = \frac{5!}{k!(5 - k)!} p^k(1 - p)^{5-k} \tag{2}$$

if $k = 0, 1, 2, 3, 4,$ or 5. (We have used the convention $0! = 1$.) This is the well-known binomial probability law, and of course $b_p(0) + b_p(1) + \cdots + b_p(5) = 1$ since this sum includes the probabilities of all possible sequences of five flips. The development leading to Equation 2, although very elementary, illustrates the computation of a consequence of a simple probability model, which typifies probability theory.

In contrast, in a statistical-inference model the chance mechanism is not completely known. Rather, the problem is to guess some of its features. A simple example is that of our coin when the value of p is unknown. The mint, it may be supposed, produces coins which, because of inhomogeneous makeup and irregular shape, range from those that almost always produce heads when flipped (p near 1) through those that produce about half heads (p near 1/2) to those that almost always produce tails (p near 0). It is not known which of these types of coin is the one we actually have, that is, which value of p between 0 and 1 characterizes the coin.

The emphasis of statistical inference is on choosing a procedure for using the observed outcome of the chance mechanism to make a guess regarding which of the *possible* mechanisms is the *actual* one at hand. In our example, this means using the observed sequence of five flips to make a guess as to the unknown value of p for our particular coin. In some reasonable sense, this guess should be as accurate as possible. However no procedure can guarantee to guess correctly. Any procedure we decide to use to make a guess about p will make the same guess in every experiment of five flips with outcome HTHTT (though perhaps some other guess whenever the outcome is TTHHH). Whatever the actual p is, the outcome HTHTT, and thus the guess obtained from it, really can arise. But it is easily seen that no meaningful statement made as this guess can be correct for all possible p. For example, if from HTHTT with two H's in five flips, we stated "I guess that p is 2/5," this would be correct if p were actually 2/5 and would be reasonably accurate if p were close to 2/5, but it would be an erroneous guess, for most practical purposes, if the coin were actually one for which $p = .9$. Similarly, any other procedure we could use can sometimes make bad quesses. As we shall see, the accuracy of different guessing procedures is a probabilistic computation.

What's the Question?

In precisely formulating a statistical problem about our coin, we begin by listing the possible relevant statements we could make about the coin. We shall decide on one of these statements after observing the outcome of the five tosses. This list can be regarded as the list of possible answers to a question about the nature of the coin.

For example, one might ask (a) "What value of p characterizes the coin at hand?" The possible answers are all real numbers between 0 and 1, and a statistical procedure for this question is a rule which associates, with each of the 32 possible sequences of five tosses, a corresponding guess as to the value of p.

A question requiring a less precise statement about p is (b) "Is the coin fair?" A statistical procedure for this question does not assert a numerical guess as to the value of p but, rather, makes one of the statements: "The coin is fair ($p = 1/2$)" or "The coin is unfair ($p \neq 1/2$)." If the mint wanted to test each coin it produced, remelting unfair ones to preclude their falling into the hands of scheming gamblers, this second question would be of interest; a more precise guess about the value of p for a given coin would be irrelevant. On the other hand, if the H and T of the coin are replaced by the life or death of a dying patient to whom a new drug is administered, then a numerical guess as to the probability p that the drug can save a life — an answer to question (a) — would often be called for.

You can think of other possible relevant questions. For example, a gambler would want to ask (c) "Is the coin fair, biased in favor of H, or biased in favor of T?" The answer tells him what use he can profitably make of a given coin: He can bet on the favored outcome in gambling games with unsuspecting victims if the coin is biased, or spend it if it is fair and, thus, has no such nefarious use. Of course, there are many similar examples of greater practical importance.

It is remarkable that, as natural as it may be to ask questions like (c), the classical development of statistics neglected such questions almost completely until Abraham Wald began his work on "statistical decision theory" in 1939 and emphasized the completely general nature of the questions one could ask. Prior to that, statisticians tended to try to push every problem into the formulation (a), called *point estimation*, or (b), called *hypothesis testing*, whether or not one of these formulations fitted the need of the customer. (A variant of (a), called *interval estimation*, was also sometimes used; here the guess took the form "I guess p to be in the interval from .37 to .45" instead of "I guess p to be .4.")

Here is a simple example of what happened when the wrong question was asked. An experimenter might be testing the productivity of several varieties of grain, wanting to decide what progress he had made in developing a new strain. The actual yield from any plant differs by a chance value, because of such effects as soil variation, from the "expected yield" for that variety, which is approximately the average yield per plant one would observe over a great many plants. The experimenter who felt he had to ask a question of type (a) or (b) would usually choose the latter and ask, "Are any of these varieties better than the standard variety most farmers now use?" If the data made him answer "No," he would return to the laboratory to try to develop other strains. But, if the answer were "Yes," he would realize he had asked the wrong question, for he would now want to know *which* of the varieties were superior ones in order to experiment further with these. He would therefore ask other questions of type (b), based on the same data: "Is variety number 1 a superior one? Are varieties 2 and 3 superior?" And so on. He would then try to combine the answers (which could even turn out to be logically incompatible) to reach a conclusion as to which varieties merited further study. By using such an *ad hoc* combination of statistical procedures which were designed to answer questions of type (b) rather than the question of real interest to him, he could end up making a rather inefficient use of the data. Moreover, he almost never knew the actual efficacy of this conglomerate procedure, that is to say, the probability that it would actually select just those varieties with expected yields appreciably greater than that of the standard variety.

Wald's approach departed from the tradition of artificially restricting attention to questions of types (a) and (b) and instead tried to ask the question of real concern. In our example that might be "Which varieties have expected yield at least 10 percent greater than that of the standard variety?" A procedure would then be designed to answer this question, with efficient use of the data and with a computation of the efficacy of the procedure in giving a correct answer.

Another interesting feature in the history of statistics is that, even in settings where a question of type (a) or (b) is appropriate, it has been less than forty years since statisticians worried much about making precise a reasonable measure of accuracy for statistical procedures and found how to construct procedures that use the data efficiently in terms of that measure. At the beginning of this century Karl Pearson and his school were the leading producers of statistical procedures, constructed largely on an intuitive basis, whose dangers we shall illustrate later in an estimation example. Beginning in 1912, R. A. Fisher showed how inefficient some of these intuitively appealing procedures could be. Over the next half century, Fisher contributed greatly to many mathematical developments in statistics and perhaps had more influence, especially on applied statisticians, than any other person. But some features of his work were unsatisfactory to those who felt that statistical inference should be based on a precise mathematical model with these two features: It makes explicit the penalties that can be incurred from reaching incorrect conclusions from the data, and it leads to the construction and use of a procedure that probably incurs small penalties by using the data efficiently. Jerzy Neyman, together with Karl Pearson's son Egon, was at the forefront of the resulting beginning of the modern mathematical theory of statistical inference.

The Cost of Being Wrong

Ideally, the precise formulation of a statistical-inference model must include not only the list of possible answers to the question asked but also a statement of the relative harm of making various incorrect answers. For example, for the question (a) of estimating the value p characterizing our coin, the loss incurred from misestimating p will presumably be larger, the further the guess is from the actual p. For our grain example of type (c), you may want to try to formulate, at least qualitatively, the way the penalty for an incorrect guess might reasonably depend on the actual average yields and the guess made. We shall simplify our subsequent calculations in this article by neglecting the precise penalty values. In particular, all incorrect guesses are regarded as equally serious in the hypothesis-testing example treated later. However, these values do, in fact, play an important part in determining which statistical procedures are good ones.

From a practical point of view, these penalty values, to be expressed in units of money or utility, are very hard even to approximate in most problems. A manufacturer may be able to assess the cost he would incur from misclassifying a defective light bulb as "good" in a quality control test. But what is the cost to an astronomer of misestimating by 20 percent the distance to a quasar he is studying? To go a step backward, in exploratory research one cannot always know in advance of the experiment the precise form of the question to be asked. Often a phenomenon shows up which has not occurred as a possibility to the experimenter. (The inviting practice of letting the data determine the question after the experiment and of answering it from these same data can lead to dangerous delusions about the accuracy of one's conclusions.) To go still another step backward in our pattern of formulation, it is often impossible to delimit precisely the class of *possible* probability mechanisms, one of which actually governs the experiment at hand.

What, then, is the worth of the theoretical developments of Neyman and Wald to the practical statistician? The answer is the same as in other areas of applied mathematics and science, where the careful study of a model that is not exactly correct can still lead to more useful decision-making or predictive procedures than will a formula or rule based on intuition or traditional rule of thumb. Moreover, the attempt to write down a precise model is often, in itself, remarkably helpful in clarifying the experimenter's thoughts about his problem. Incidentally, in view of the possible inaccuracy of the model, an important topic is the study of properties of a statistical procedure designed for use with one class of probability mechanisms when, because of incorrect formulation of the model, the actual mechanism is outside that class. This is beyond the realm of the present article.

Two simple examples illustrate some of the ideas arising earlier. In the first of these, we shall see that, even in the simplest models for a question of type (b), there are many procedures which use the data efficiently, and the choice among them is usually difficult.

Testing Between Simple Hypotheses

We shall continue to use the language of independent tosses of a coin, although you can, of course, imagine H and T to stand for the outcomes of any dichotomous experiment. Suppose our coin is known in advance to be characterized by either the value $p = 1/3$ or by the value $p = 2/3$; no other values are possible. On the basis of three independent flips, we are to guess which value of p actually characterizes the coin. Each of the two possible guesses specifies a single probability mechanism, unlike the example "The coin is unfair ($p \neq 1/2$)" discussed earlier, where the guess did not attempt to describe the exact value of p. When a "hypothesis" consists of just one possible mechanism, it is said to be "simple." Thus, statisticians refer to our present problem as one of testing between two simple hypotheses. The characterization of procedures that use the data efficiently will turn out to be particularly simple for such problems and much simpler than that for other hypothesis-testing problems.

Here are five possible prescriptions for making a guess from the data; you can easily write down others.

Procedure 1: guess "$p = 2/3$" if there are at least two H's in the three tosses; guess "$p = 1/3$" otherwise.

Procedure 2: guess "$p = 2/3$" if the first H precedes the first T or if no T occurs; guess "$p = 1/3$" otherwise.

Procedure 3: guess "$p = 2/3$" if there is at least one H in the three tosses; guess "$p = 1/3$" otherwise.

Procedure 4: guess "$p = 2/3$" if there is at most one H in the three tosses; guess "$p = 1/3$" otherwise.

Procedure 5: ignore the data and always guess "$p = 2/3$."

Which of these procedures would you use? Perhaps many people would feel intuitively that there is more tendency to obtain H's when $p = 2/3$ than when $p = 1/3$, so that Procedures 1, 2, and 3 do not seem too unreasonable and Procedures 4 and 5 seem unreasonable (the former because it works in the wrong direction; the latter, because it makes no use of the data). For a precise analysis, we first use a calculation like that of Equation 1, near the beginning of this essay, to list the probabilities of the eight possible sequences of three tosses under each of the two possible probability mechanisms (see Table 1).

Table 1

Outcome	Probability of Outcome when $p = 1/3$	Probability of Outcome when $p = 2/3$
HHH	1/27	8/27
HHT	2/27	4/27
HTH	2/27	4/27
THH	2/27	4/27
HTT	4/27	2/27
THT	4/27	2/27
TTH	4/27	2/27
TTT	8/27	1/27

Then, for each procedure, we use Table 1 to compute the probability that the chance outcome will be such as to lead to a correct guess when $p = 1/3$. Next, we perform the same computation with $p = 2/3$. For example, when $p = 1/3$, Procedure 2 makes the correct guess "$p = 1/3$" if there is a T on the first toss, that is, if the outcome is THH or THT or TTH or TTT, with a total probability of 18/27, computed from the middle column. Similarly, summing the probabilities in the last column for the four other possible outcomes which lead to the guess "$p = 2/3$" yields 18/27 for the probability of a correct guess when $p = 2/3$. A similar calculation for the other procedures gives Table 2.

What does this table show us about the appeal of various procedures? If a procedure were perfect, it would have probability one (certainty) of making a correct guess, whether the actual p is 1/3 or 2/3. It is not hard to see that no such procedure exists. Procedure 5 is always correct if $p = 2/3$, but never right if $p = 1/3$. The procedure which ignores the data and

Table 2

Procedure	Probability of Correct Guess when $p = 1/3$	Probability of Correct Guess when $p = 2/3$
1	20/27	20/27
2	18/27	18/27
3	8/27	26/27
4	7/27	7/27
5	0	1

always guesses "$p = 1/3$" would have the opposite behavior. No other procedure has probability one of making a correct guess, whatever p may be.

Moreover, we see that no procedure maximizes the probability of making a correct guess when $p = 1/3$ and also when $p = 2/3$: Procedure 5, alone among all procedures, maximizes the probability of making a correct guess if $p = 2/3$, but it minimizes this probability when $p = 1/3$. Since a perfect maximizing procedure does not exist, how are we to select the procedure to be used?

To start with, Procedures 2 and 4 can be eliminated from contention since, whatever the actual value of p, each of these has a smaller probability of making a correct guess than does Procedure 1. Thus, while we have not yet decided on which procedure to use, we can tell anyone who proposes to use Procedure 2 that he should not do so, since we can give him another procedure which has a higher probability of yielding a correct guess, whether the actual p is 1/3 or 2/3. A subject of considerable study among theoretical statisticians is the characterization in different problems of those procedures, called "admissible," which remain after the inferior procedures, such as Procedures 2 and 4, have been eliminated. The choice of the procedure to be used is thereafter restricted to the admissible procedures.

Procedures 1, 3, and 5 can be shown to be among the admissible procedures in our example; you can see for yourself that none of these three eliminates either of the others by having a higher probability in both columns.[2] The choice among Procedures 1, 3, 5, and the other admissible procedures must now involve some additional criterion which we have not yet mentioned. There are philosophical differences among statisticians as to the appropriateness of various criteria which have been suggested and used over the years. I shall next mention a few of these criteria.

[2] Each admissible procedure is characterized, by the Neyman–Pearson Lemma, to guess "p = 2/3" for those outcomes in Table 1 for which the ratio of the last to the middle column is greater than some constant c. You will see that c can be taken to be 1 for Procedure 1, 1/4 for Procedure 3, and 1/10 for Procedure 5. For technical reasons, theoretical statisticians also include procedures which, when the outcome yields a ratio c, perform an auxiliary experiment to decide which guess to make. These need not concern us here.

Basic Ways of Choosing

The approach of Fisher in hypothesis testing, usually in more complex examples than ours, was to use a procedure that attained a previously specified probability of making a correct guess under one particular probability mechanism. This mechanism was often an older theory whose validity was to be tested by the experiment or "no difference between new varieties and old" in the experiment we discussed in leading up to question (c). In our present example, let us suppose that $p = 1/3$ is the chosen mechanism. We must then specify the value a procedure is to yield in the middle column of Table 2. The Neyman–Pearson theory added to this formulation an aspect never treated by Fisher: Among all procedures with the specified value in the middle column, one should select that procedure with largest entry in the last column. Thus, among all tests with 8/27 in the second column, of which there are many in addition to Procedure 3 (for example, guess "$p = 1/3$" for outcomes THT and TTH, and "$p = 2/3$" otherwise), the Neyman–Pearson Lemma characterizes Procedure 3 as the one with maximum probability of making a correct guess when $p = 2/3$. If one accepts the figure 8/27, there is no doubt about using Procedure 3. The practical shortcoming is: Why 8/27 rather than some other figure (for example 20/27, which would have dictated using Procedure 1)? The value has often been chosen by tradition alone, in various ways in different fields of applications.

A second criterion is given by the "minimax" principle. For each procedure, we compute the minimum of the two probabilities listed in Table 2: 20/27 for Procedure 1, 8/27 for Procedure 3, 0 for Procedure 5. This figure gives a measure of the worst possible performance of the procedure. The pessimist, worried about the possibility of encountering a coin whose p yields this worst performance, may want to choose a procedure for which this minimum probability is as large as possible. In the present example, that is Procedure 1. There are also some objections to this approach, but space does not permit us to discuss them here.

A third possibility is to compute for each procedure some weighted average of the two probabilities listed in Table 2 and to choose a procedure that maximizes this average. For example, if we compute $\frac{1}{2}$ the middle column plus $\frac{1}{2}$ the last column for each procedure, we find that this average is largest for Procedure 1. On the other hand, $\frac{1}{4}$ of the middle column plus $\frac{3}{4}$ of the last column is largest for Procedure 3. Thus, different weighted averages favor the choice of different procedures. The practical difficulty here is: How do you choose the weights? In a very few problems, we know so-called *prior* probabilities that the actual physical mechanism will be of one form or another. For example, in our coin problem, we might know that, over a long period, $\frac{3}{4}$ of the coins have come out of the mint with $p = 2/3$, and $\frac{1}{4}$ with $p = 1/3$. In such a case, it is appropriate to use the prior probabilities $\frac{3}{4}$ and $\frac{1}{4}$ as weights, since the resulting average our procedure will maximize is then the total probability of a correct guess. This yields the choice of Procedure 3 in the example just given. Because the general scheme for computing such procedures uses the simple probabilistic formula known as Bayes' Theorem, the resulting procedure is called a Bayes procedure for the given prior probabilities. But there are few practical examples where such prior probabilities are known. In the absence of such knowledge, it is tempting to use equal prior probabilities to represent one's ignorance (which would lead to the use of Procedure 1 in our example) or to use the recently much-publicized subjectivist approach, which employs, in place of unknown physical prior probabilities, corresponding "subjective probabilities," which are supposed to reflect a quantitative measure of the customer's feelings about the problem. Both these possibilities have also received considerable criticism, which we cannot discuss here.

A fourth possibility is to take note of the symmetry of the problem, as reflected in the fact that relabeling H as T and vice versa means interchanging the values $p = 1/3$ and $p = 2/3$, since a coin with probability 1/3 of yielding H has probability 2/3 of yielding T (which becomes H under the relabeling). The symmetry of the problem suggests the criterion of using a symmetric procedure, one that guesses "$p = 1/3$" for a given sequence (for example, TTH) if and only if it guesses "$p = 2/3$" for the relabeled sequence (for example, HHT). In our example, only Procedure 1, among all admissible procedures, has this symmetry; in fact, it can be shown to be better than all other symmetric procedures. Thus, under the criterion of symmetry (usually called "invariance"), we would use Procedure 1. The difficulty with using this approach is that there are more complex problems in which no invariant procedure is admissible, so that the invariance criterion cannot lead to the choice of a satisfactory procedure; and there are other problems lacking symmetry, so that all procedures are equally "symmetric," and this criterion does not choose among them. An example of the latter is our coin problem with the two possible values $p = 1/3$ and $p = 2/3$ replaced by $p = 1/5$ and $p = 2/3$. Under relabeling, these become $p = 4/5$ and $p = 1/3$, which are not merely the original pair of values in reversed order. There is no symmetry to the problem, so all procedures are equally symmetric.

You may well wonder at this point which procedure to use in our example. I cannot tell you without further knowledge of the background of the problem, the real meaning of the two events we labeled H and T (they might mean "life" and "death," etc.), the use you will make of your guess, and a re-examination of the relative losses (which we have tacitly assumed equal) for the two possible types of incorrect guess. Even with this knowledge, I might find it difficult to tell you that a particular procedure is the one and only obvious one to use, although there are many circumstances where I would stop this hedging and use Procedure 1.

I have included such a long description of some of the criteria people use to select a procedure and have indicated that all of them have shortcomings, in order to emphasize the view of many theoretical statisticians — that there is no simple recipe which will tell you how to choose a statistical procedure in all possible settings. This is a frustrating aspect of the subject and perhaps one without parallel in other mathematical areas. It presents a large target for future theoretical work. At the same time it points up the danger of using some simple, superficially appealing recipe to select a statistical procedure, and the need for expertise in that selection.

An even more fundamental danger occurs in the use of an intuitively appealing rule of thumb to construct a procedure, without any consideration of admissibility. The statistical literature is full of suggestions of procedures like Procedure 2 and is almost devoid of mention of Procedure 5 (which should be used if, for example, one knows that the prior probability that $p = 1/3$ is quite small, say $1/10$). In other settings where the mechanism is nothing as simple as coin-flipping, it is even easier to be led astray by intuition. We shall illustrate this now in an estimation problem that will exemplify the Pearson–Fisher controversy mentioned earlier.

An Estimation Example

A particle is moved along some fixed line by a force field around it. In each of three independent experiments, the particle, at rest, is placed at the same point P, and its position is measured one second later. Taking P as the origin, if the field were unaltered by outside influences and if the measurements were without error, the measured position of the particle after one second would be the same value, say θ millimeters, in each experiment. However, the experiments are not perfect, and we suppose that in each experiment the probability that the recorded measurement will be θ, $\theta - 1$, or $\theta + 1$ is $1/3$ for each. (This law of errors is being chosen for arithmetic simplicity rather than physical reasonableness, but it will illustrate a phenomenon that could also exist for more reasonable but more complicated models.) Here the force field, and thus θ, is unknown in advance of experiments, the object of which is to "guess the value of θ." Let us call the three measurements X_1, X_2, X_3. A statistical procedure is a real-valued function of these three quantities, that is, a rule for computing from them a real number that will be stated as our guess as to the actual value of θ for the particular field at hand.

A common intuitive line of reasoning for selecting a procedure begins by remarking that the value θ is the mean, or first moment, of the mass distribution corresponding to our law of errors, which assigns mass $\frac{1}{3}$ to each of the points $\theta - 1$, θ and $\theta + 1$. This reasoning then proceeds to note that we can summarize the three measurements in the form of an "empiric mass distribution" that assigns to each real value the proportion of observations taking on that value. (For example, if $X_1 = 16.2$, $X_2 = 17.2$, $X_3 = 16.2$, this empiric mass distribution would assign mass $\frac{2}{3}$ to the value 16.2 and $\frac{1}{3}$ to the value 17.2.) Since this chance empiric mass distribution has some tendency to resemble the underlying mass distribution corresponding to the probability law, the reasoning concludes that a guess as to the mean θ of the latter should be obtained from the mean of the former, which is easily seen to be the "sample mean" $(X_1 + X_2 + X_3)/3$. This is a simple example of Pearson's "method of moments."

We can easily improve upon this procedure by altering the guessing rule whenever two of the three X_i take on a value 2 mm away from the third, guessing the midpoint between these two values rather than the sample mean in such cases. The sample mean will be in error by $\frac{1}{3}$ mm for such outcomes, while the new guess will be errorless. For example, if $X_1 = 14$, $X_2 = 16$, $X_3 = 14$, you can see that θ must be 15 because of the particular form of our law of errors, and this is what our new procedure guesses it to be. But the sample mean yields a guess of $\frac{44}{3}$, underestimating θ by $\frac{1}{3}$. Since the two procedures yield the same guess (and, hence, the same error) for all other types of outcomes, the new rule is certainly preferable to the sample mean, and it is not hard to see that the probability is $\frac{6}{27}$ that the outcomes will be of the type where this reduction of error occurs.

You may well think it an obvious improvement on the sample mean to replace it by the improved guess in the foregoing situations where the exact value of θ is obvious from the outcomes. But it is easy to alter the law of errors slightly, at the expense of arithmetic simplicity, to obtain a model where the sample mean can be improved upon without such an obvious motivation. The point is that the intuitively appealing guess, "sample mean," used blindly in so many experimental settings, may be appropriate for some assumed laws of errors and terribly inefficient for others. Only a precise probabilistic analysis, and not any intuitive rule of thumb, can determine what procedures are reasonable ones for a given statistical model.

That is what the modern mathematical theory of statistical inference is all about.

BIBLIOGRAPHY

Herman Chernoff and Lincoln E. Moses, *Elementary Decision Theory*, Wiley, New York, 1959.

J. L. Hodges, Jr., and E. L. Lehmann, *Basic Concepts of Probability and Statistics*, Holden-Day, San Francisco, 1964.

BIOGRAPHICAL NOTE

Jack Kiefer has spent most of his professional career at Cornell University, where he is now a professor of mathematics. As an undergraduate at Massachusetts Institute of Technology, he concentrated on engineering and economics until he became interested in mathematical statistics. After taking his S.B. and S.M. degrees there, he did graduate work and received his Ph.D. at Columbia University in 1952. His research has been primarily in statistics and probability theory. He has received a Guggenheim Fellowship.

When a mathematician comes upon an exciting new idea, he usually tries to generalize it as far as possible. The familiar Pythagorean theorem, which applies to right triangles in plane geometry, for example, has been extended to spaces of many dimensions, even infinitely many dimensions. As so often happens with such mathematical flights of fancy, this generalization turns out to be useful in many branches of physics and engineering. Recent investigations are carrying the generalization still further, to infinite-dimensional curved spaces "glued together" in complicated ways.

Functional Analysis

J. T. Schwartz

Mathematical research *is* the invention of fruitful new ideas. Its practice shows, however, that really good new ideas are hard to come by. This explains the mathematician's tendency to attempt the generalization of every idea. By generalizing an idea, one can carry over the insight it contains from its area of origin to other areas, perhaps even to areas that, at first glance, appear unrelated. Only through generalization can an idea's full scope be revealed. The successive stages of a successful generalization will at times extend over years or decades and may embody the work of many mathematicians, each of whom succeeds in perceiving the core idea of the generalization in a less limited way than his predecessor. Thus, what started as a limited idea can grow and spread and unify important concepts in many diverse fields.

One of the first great ideas of mathematics, the theorem of Pythagoras, provides a particularly elegant example of the mathematical process of generalization. As everyone will recall from high school mathematics, this theorem states that the length of the diagonal of a rectangle, squared, is equal to the sum of the squares of the lengths of its two sides (see Figure 1). We express this in the familiar algebraic terms $L^2 = A^2 + B^2$, where A and B are the lengths of the two sides of a rectangle, and L is the length of its diagonal. This ancient formula (and here begins a major develop-

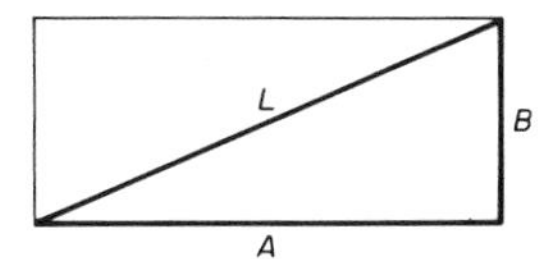

FIGURE 1. *Pythagoras' theorem in the plane.*

ment) generalizes at once from the plane to three-dimensional space. If L is the length of the diagonal of a rectangular box, and A, B, and C are the sides of the box, then $L^2 = A^2 + B^2 + C^2$. The proof is immediate: Apply the plane Pythagoras theorem to the rectangle of B and C, to obtain $(L')^2 = (L'')^2 = B^2 + C^2$ for the diagonal of that rectangle: Then apply the plane Pythagoras theorem once more, this time to obtain the rectangle with sides A, L', A', and L'' (see Figure 2), to obtain

$$L^2 = A^2 + (L')^2 = A^2 + B^2 + C^2.$$

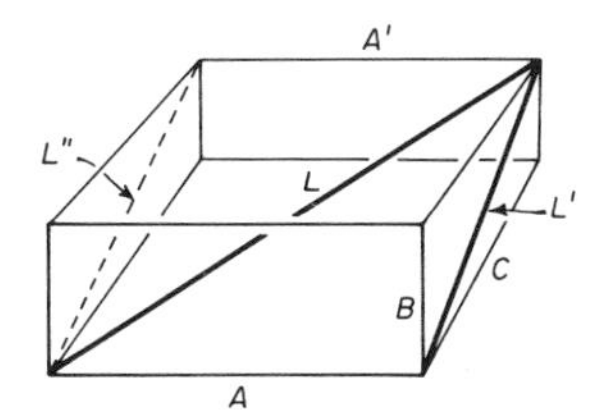

FIGURE 2. *Pythagoras' theorem in three-dimensional space.*

All this was known at the time of Euclid. Early in the modern period, the invention of coordinate geometry by Descartes suggested a still more far-reaching generalization of Pythagoras' theorem. According to Descartes, a point in the plane can be located by giving its two position coordinates (x, y), which, geometrically speaking, are the projections of the point along two perpendicular axes. Similarly, we can locate a point in three-dimensional space by giving its three position coordinates (x, y, z); geometrically speaking, these are the projections of the point on three perpendicular axes in three-dimensional space (see Figure 3). Having introduced these coordinates, we may resolve, following Descartes, always to deal with points p in space in terms of their three coordinate numbers (x, y, z). But then how is the fundamental quantity of geometry, the distance between p and another point q, to be expressed? This question is

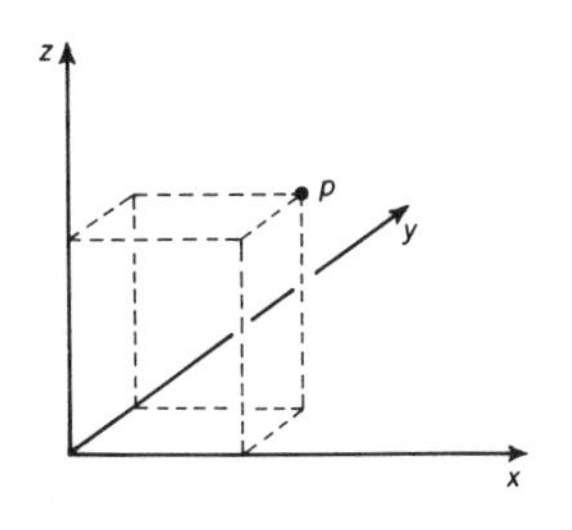

FIGURE 3. *The three coordinates (x, y, z) or a point p in three dimensional space.*

answered at once by Pythagoras' theorem. If the point p has the coordinate numbers (x, y, z) and if the point q has the coordinate numbers (u, v, w), then, Pythagoras' theorem tells us, the square of the distance between p and q is the sum of the squares of the differences of their coordinate numbers. That is, the distance from p to q, squared, is

$$(x - u)^2 + (y - v)^2 + (z - w)^2.$$

Thus, Pythagoras' theorem emerges as the basic principle through which the geometric notion of distance can be harmoniously related to the more algebraic idea of coordinate numbers introduced by Descartes. When one has come this far, considerably more extensive generalizations of Pythagoras' theorem come to the surface. If any two numerical coordinates (x, y) determine a point p in the two-dimensional plane and if any three numerical coordinates (x, y, z) determine a point p in three-dimensional space, is it not reasonable to consider that any four numerical coordinates (x, y, z, t) will determine a point p in four-dimensional space, that any five numerical coordinates (x, y, z, t, s) will determine a point p in five-dimensional space, etc.? If, however, we attempt to define spaces of four, five, and more dimensions in this way, how can we express the distance between points in these spaces? The Pythagorean theorem supplies the key: In strict analogy with the two- and three-dimensional formulas that follow from this theorem, we take

$$\text{Distance from } p \text{ to } q\text{, squared,} = (x - u)^2 + (y - v)^2 + \cdots + (t - r)^2$$

to define the distance between a point p with coordinate numbers $(x, y, z, \cdots, t)$ and a point q with coordinate numbers $(u, v, w, \cdots, r)$ in four, five, or more dimensions. In this way, Pythagoras' theorem emerges as the foundation not only of plane geometry and of the geometry of three-dimensional space but of the geometry of space of four and more dimensions.

Infinitely Many Dimensions

As observed by the German mathematician David Hilbert early in the present century, we can extend Pythagoras' theorem still further. A point p in a space of any finite number n of dimensions (n-dimensional space) is, as we have seen, determined by its n numerical coordinates. Suppose that we enumerate these coordinates, from first to last, as $(x_1, x_2, x_3, \cdots, x_n)$. Why, then, must the total collection of coordinates be finite? We can, in fact, introduce an *infinite-dimensional* space, whose points p are determined by unending sequences $(x_1, x_2, x_3, \cdots, x_n, x_{n+1}, x_{n+2}, \cdots)$ of coordinate numbers. If q is a second point in this infinite-dimensional space, or "Hilbert space," whose sequence of coordinate-numbers is

$$(y_1, y_2, y_3, \cdots, y_n, y_{n+1}, y_{n+2}, \cdots),$$

we can define the distance between p and q by the following evident generalization of Pythagoras' formula:

Distance from p to q, squared,

$$= (x_1 - y_1)^2 + (x_2 - y_2)^2 + (x_3 - y_3)^2 + \cdots + (x_n - y_n)^2 + (x_{n+1} - y_{n+1})^2 + \cdots,$$

the sum being indefinitely extended to a (convergent) ultimate total. Thus, in addition to plane, three-dimensional, four-dimensional, etc., geometry, we may even consider an *infinite-dimensional geometry.*

An even more far-reaching generalization is possible. Any sequence of numbers and, in particular, the sequence $(x_1,x_2,x_3,\cdots)$ of coordinate-numbers defining a point in Hilbert space of infinitely many dimensions can be regarded as defining a *rule* or *function* which assigns to every whole number n a decimal number x_n. If we picture the sequence of numbers x_n by a sequence of blocks of unit width (see Figure 4), the sum of all the

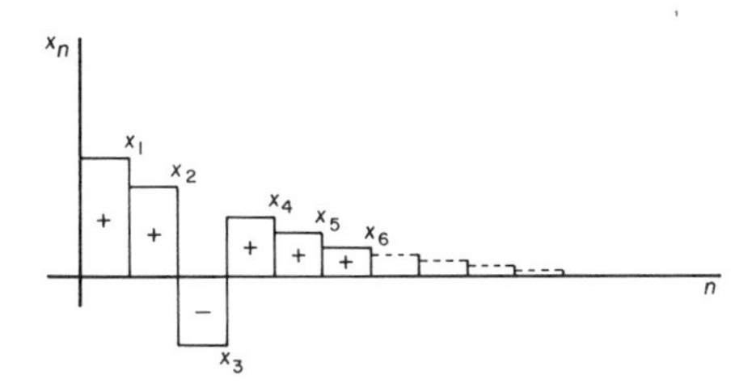

FIGURE 4. *Block graph of a sequence* $(x_1, x_2, x_3, \ldots)$ *of numbers. (The total area of the blocks represents the sum of the terms in the sequence.)*

terms in the sequence equals the total area of all the blocks in the picture. This suggests that, instead of discrete sequences of numbers $(x_1,x_2,x_3,\cdots)$, we consider functions $x(t)$ that assign a decimal number x to every value of t in a continuous range; let us say (for the sake of definiteness) to all positive numbers t. We may then regard the total area under the graph of such a function $x(t)$, its so-called *integral*, as a replacement for the sum $x_1 + x_2 + x_3 + \cdots$ of a discrete sequence of values (cf. Figure 5).

In analogy with Hilbert's *discretely* infinite-dimensional space, we may

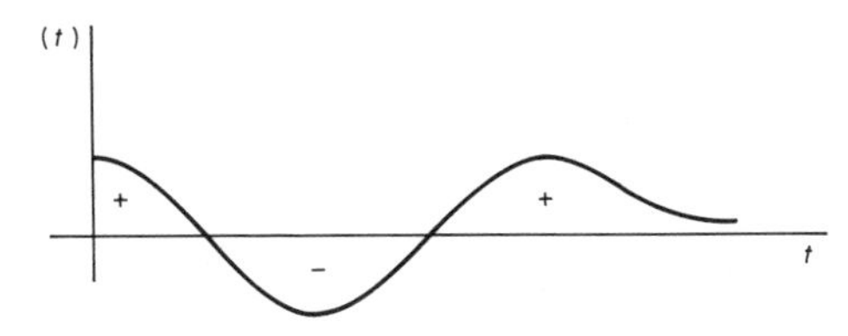

FIGURE 5. *Continuous graph of a smooth function* $x(t)$ *of a positive variable* t. *(The total area between the curve and the* t*-axis is the* integral *of the function.)*

then construct a *continuously* infinite-dimensional space. Its points p are defined by functions $x(t)$ which assign a decimal value to each number in a continuous range. If q is a second such point, defined by another function $y(t)$ of the variable t, we may define the distance between p and q by the following generalization of Pythagoras' formula, the most far-reaching of all the generalizations of this formula we have described:

Distance from p to q, squared,
= area under the graph of the function $(x(t) - y(t))^2$.

The steps outlined thus far have carried us, through a progressive sequence of generalizations, from Pythagoras' formula for plane figures to a version of Pythagoras' formula which applies to spaces of infinitely many dimensions. The individual elements in these infinite-dimensional spaces are, in fact, functions. Such infinite-dimensional spaces are called *function spaces;* they give their name to the study of their properties, called *functional analysis.*

Marriage of Geometry with Function Theory

In the process of arriving at these continuously infinite-dimensional spaces, we have penetrated to a level of generality at which geometry and function theory fuse and in which well-known facts of two- and three-dimensional geometry can lead, by easy analogy, to deep statements about functions of a continuous parameter. Let us examine some of the most fruitful of these geometric analogies.

1. In three-dimensional geometry, the *norm* of a coordinate point p, given by a set (x_1, x_2, x_3) of coordinate numbers, is the distance between the point p and the *coordinate origin,* whose three coordinates are (0, 0, 0). Thus, the square of the norm of p is $x_1^2 + x_2^2 + x_3^2$. In functional geometry, the analogous norm is the area under the graph of the function $(x(t))^2$. Now, a norm of this sort occurs independently in many branches of physics and engineering. In electrical engineering, the power dissipated in an electrical circuit can be calculated from the norm of a voltage function. In electromagnetic theory, the norm of a field-intensity function gives the total energy of an electromagnetic wave. In acoustic theory, the norm of a pressure function gives the total energy of a sound wave. Since such basic quantities in each of these three physical theories are formally identical with a basic geometric quantity in function-space geometry, we may readily surmise that the methods of functional analysis will prove highly applicable in all these theories; this, indeed, proves to be the case.

2. If p and q are points in the plane, with coordinate numbers (x_1, x_2) and (y_1, y_2), respectively, and if 0 denotes the origin of coordinates, then, according to Pythagoras' theorem, the lines $\overset{\frown}{op}$ and $\overset{\frown}{oq}$ are perpendicular if and only if the square of the distance from p to q is equal to the sum of the two squared distances $(\overline{op})^2$ and $(\overline{oq})^2$ (cf. Figure 6). That is, the points

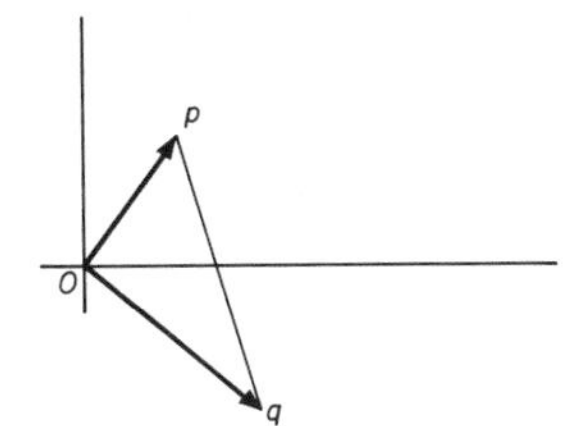

FIGURE 6. *Distance relations along perpendicular lines.*

with coordinates (x_1, x_2) and (y_1, y_2) are along perpendicular lines through the origin of coordinates if and only if the relation

$$(x_1^2 + x_2^2) + (y_1^2 + y_2^2) = (x_1 - y_1)^2 + (x_2 - y_2)^2$$

holds. This relation may readily be simplified algebraically to the equivalent condition

$$x_1y_1 + x_2y_2 = 0.$$

The same reasoning, applied to a pair of points, p and q, in three-dimensional space with coordinates (x_1, x_2, x_3) and (y_1, y_2, y_3), respectively, would show that they lie along perpendicular lines through the coordinate origin if and only if

$$x_1y_1 + x_2y_2 + x_3y_3 = 0.$$

Under these conditions, we say that the points p and q in three-dimensional space (or in the plane) are *orthogonal*. This geometric notion extends without difficulty to all of the spaces of dimension higher than three, described earlier. We may say that

(a) Two points, p and q, in n-dimensional space, with coordinates $(x_1, x_2, \cdots, x_n)$ and $(y_1, y_2, \cdots, y_n)$, respectively, are orthogonal if

$$x_1y_1 + x_2y_2 + \cdots + x_ny_n = 0.$$

(b) Two points p and q in the infinite-dimensional space of sequences, with infinite coordinate sequences $(x_1, x_2, \cdots)$ and $(y_1, y_2, \cdots)$, are orthogonal if

$$x_1y_1 + x_2y_2 + \cdots = 0.$$

(c) Two points, p and q, in the infinite-dimensional space of functions, which correspond to functions $x(t)$ and $y(t)$, are orthogonal if and only if the total area under the graph of the product function $x(t)y(t)$ is zero.

In the plane, the largest possible number of mutually perpendicular lines is two; in three-dimensional space, three. These elementary geometric facts at once suggest corresponding statements for the higher-dimensional spaces we have considered. We may guess, for example, that the largest possible number of mutually orthogonal points in four-dimensional space is four, in five-dimensional space is five, in 100-dimensional space is 100, in n-dimensional space is n. All of these guesses turn out to be true. We may guess that both the infinite-dimensional space of sequences and the infinite-dimensional space of functions contain infinite numbers of mutually orthogonal points. These guesses also turn out to be true.

A *rigid motion* in the plane or in three-dimensional space is a motion of the plane or the three-dimensional space over itself which does not change the distance between points. In three-dimensional space, any set of three mutually perpendicular lines may be moved onto any other set of three mutually perpendicular lines by a rigid motion of space; a similar, even simpler, statement holds for the plane. We may therefore guess that, in four-dimensional space, any set of four mutually perpendicular lines may be moved onto any other set of four mutually perpendicular lines by a rigid motion of space. This guess turns out to be true. We find a common generalization in the statement that, in a space of any number of dimensions, finite or infinite, any set of perpendicular lines *containing as many mutually perpendicular lines as is possible* may be moved onto any other such set of lines by a rigid motion of the whole space. The truth of this statement shows that all our n-dimensional and infinite-dimensional spaces are as homogeneous and as undifferentiated as to direction as is the Euclidean plane.

3. In the plane, any set of two perpendicular lines may serve as coordinate axes. In three-dimensional space, any set of three mutually perpendicular lines may serve as coordinate axes. We have seen above that in n-dimensional space, any set of n mutually perpendicular lines may serve as coordinate axes.

Two Basic Abstractions

Two useful abstract principles underlie all that has been said above about spaces of many dimensions, infinite-dimensional sequence spaces, and spaces of functions. The first abstract principle is that any geometric statement which pertains both to the plane and to three-dimensional space *may* be sufficiently general to extend successively to four, five, and n-dimensional space, and thence to function space. Any such geometric statement may therefore lead to a principle useful for the study of functions, equations involving functions, and so on.

The second abstraction is even broader and is useful not only in functional analysis but in mathematics very generally. It may be put as follows: In geometry, we do not study single points, but regard each point as being a particular member of an entire universe of points (the plane, three-dimensional space, four-dimensional space, and so on, as the case may be). That is, in geometry, we study points not in themselves but in context. In the same way, we see that individual sequences and functions may be considered not only in themselves but in context as particular elements of whole spaces of sequences or whole spaces of functions. The significance of a particular property of an individual function, a property that may be quite surprising when the function is considered individually, may be obvious when the function is considered in the larger context of a space of functions. Use of context in this way as a principle of explanation and of investigation is a very powerful mathematical device. The principle of context reaches one of its fullest expressions within functional analysis in the treatment of function spaces. Such use of context is, however, very classical in mathematics. Thus, in optics one analyzes the particular path of a light ray through an optical medium as that path among all possible paths between the end points of the ray which the light can traverse in minimum time. In electromagnetic theory, one studies each actual distribution of electricity on a conducting material as that distribution with the least energy of distributions. Similarly, the actual shape of soap films constructed on a wire frame is the particular shape, among all shapes, which minimizes the area of the film. Innumerable examples may be cited from almost every domain of physics. Functional analysis unifies all these examples, putting them all into the geometric framework of function space.

We may profitably consider still another embodiment of these abstractions. Again we shall begin with a geometric statement about two- and three-dimensional spaces, drawn this time not like the Pythagorean theorem from ancient geometry but from the more modern "rubber sheet" geometry known as topology. Consider a disc in the plane and a mapping of this disc into itself, for example, a rule which assigns to every point in the disc some other point of the disc, or possibly itself (cf. Figure 7). The fa-

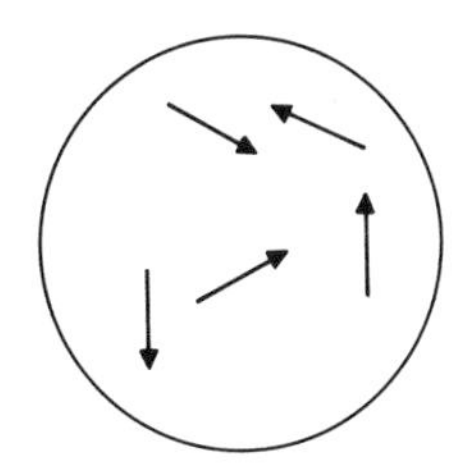

FIGURE 7. *A mapping of a disc in the plane into itself.*

mous "fixed point theorem" of Brouwer states that, if the mapping is *continuous* (if points near to each other in the disc are always sent into points near to each other), there must be at least one point of the disc which is not moved at all. Consideration of a few examples will convince us of this surprising inevitability. For example, if the disc is rotated about its center, every point but the center moves — but the center does not move. If the disc is shrunk down to a single one of its points, the point to which the disc is shrunk does not move. If the disc is reflected across an axis, none of the points of the axis moves. The fixed-point theorem tells us that, no matter how we arrange our mapping, there must always be at least one point that does not move. A similar statement holds not only for a disc in the plane but for a solid ball in three-dimensional space and for corresponding figures in four-, five-, and higher-dimensional spaces. In view of this analogy between finite dimensional spaces and function spaces, one may suspect immediately that a similar statement also holds for spaces of functions. Such a statement, embodying the so-called fixed-point principle for Hilbert's infinite-dimensional space, was in fact guessed at and proved by the Polish mathematician J. Schauder in 1930.

The fixed-point principle is, by virtue of its very great generality, applicable to the most varied situations. We may see this most easily by reformulating the principle slightly. Let p be a point in the space of n-dimensions, and let its coordinate numbers be $(x_1, x_2, \cdots, x_n)$. The *unit ball* in n-dimensional space is the collection of all points whose distance from the origin does not exceed 1. Pythagoras' theorem enables us to express this geometric condition in terms of the coordinates of the point by the condition that $x_1^2 + x_2^2 + \cdots + x_n^2$ does not exceed 1. A continuous mapping of the ball into itself is determined by *any* set of formulas which determines n new coordinates $(y_1, \cdots, y_n)$ in terms of the n original coordinates $(x_1, \cdots, x_n)$ in such a fashion that the inequality

$$x_1^2 + \cdots + x_n^2 \leq 1 \qquad \text{(sum of } x_j^2 \text{ less than or equal to 1)}$$

always implies the inequality

$$y_1^2 + \cdots + y_n^2 \leq 1.$$

If such is the case, the fixed-point theorem tells us that, irrespective of the particular formulas by which the coordinates $(y_1, \cdots, y_n)$ are determined in terms of the $(x_1, \cdots, x_n)$, there must exist at least one point p for which the original coordinates $(x_1, \cdots, x_n)$ and the "mapped" coordinates $(y_1, \cdots, y_n)$ are identical. If $y_j = y_j(x_1, \cdots, x_n)$ designates the particular formulas that express the y's in terms of the x's, this means that the set of equations

$$\begin{aligned} x_1 &= y_1(x_1, \cdots, x_n), \\ x_2 &= y_2(x_1, \cdots, x_n), \\ &\vdots \\ x_n &= y_n(x_1, \cdots, x_n) \end{aligned}$$

always has a solution. We emphasize again that we may make this assertion without knowing anything at all about the particular form of the formulas $y_j(x_1, \cdots, x_n)$.

In Hilbert's infinite-dimensional function space, we may still define the unit ball as the set of all points p whose distance from the zero function (origin of coordinates) does not exceed 1. Such points correspond to functions $x(t)$ for which the area under the graph of $(x(t))^2$ does not exceed 1. Then, if we take any (suitably continuous) formula $y(x(t))$ through which a new function $y(t) = Y(x(t))$ is determined from the initial function $x(t)$ and if the area under the graph of $(y(t))^2$ is always less than 1, provided that the same condition applies to the graph of $(x(t))^2$, it follows that the equation $x(t) = Y(x(t))$ has a solution. We repeat that this conclusion follows on general geometric grounds, irrespective of any complexities inherent in the formulas Y.

To take an example, we state that the fixed-point principle immediately implies the existence, for all sufficiently small numbers a, of a solution of the complicated integral equation

$$x(t) = a\left(\int_0^t \cos(sx(s))(x(s)^2 + 3x(s))\,ds\right)^{1/2}$$

To verify this assertion, the mathematician need not make a detailed analysis of the equation but need merely recognize it as an equation to which the general geometric fixed-point principle will apply. Here we observe an overwhelming "force of context": The existence of a solution of the equation above is, as it were, a matter of context and context alone.

The fixed-point principle has proved, by virtue of its generality, to be applicable in fluid mechanics, the theory of elastic bodies, the theory of periodic motions of particles in accelerators, and many other fields. In fact, if an equation for a function is complicated and nonlinear and if the existence of a solution for it cannot be shown easily by the fixed-point principle, to show that the equation can be solved at all may involve a very major mathematical effort.

Curved Spaces

Until now we have considered only examples which have their intellectual kernel in geometric truths concerning flat two- and three-dimensional space, that is, in results concerning the ordinary plane and the three-dimensional space of Euclid. Everyone knows, however, that the plane is not the only possible surface, that curved surfaces also exist. Corresponding "curved manifolds" of functions, the infinite-dimensional functional analogues of curved two-dimensional surfaces, have been studied for some time; they have been the object of particularly fruitful investigations in the past few years. The most rewarding properties of finite-dimensional spaces, for the purpose of infinite-dimensional functional generalization, have not been those that pertain to the particular shapes of the spaces but those deeper properties that pertain to the over-all manner in which a space is "glued together" in terms of its parts: the so-called *topological* properties of the space. Such topological properties, for example, distinguish a torus (the surface of a ring) from the surface of a sphere: the torus

FIGURE 8. *Torus and sphere.*

"has a hole," the sphere does not (cf. Figure 8). This fundamental distinction between torus and sphere has many important consequences for the over-all behavior of families of curves on the two surfaces and also for the behavior of functions defined on the two surfaces. Thus, for instance, the torus can be covered by a smooth family of nonintersecting curves (cf. Figure 9); the sphere cannot, as is suggested by the existence

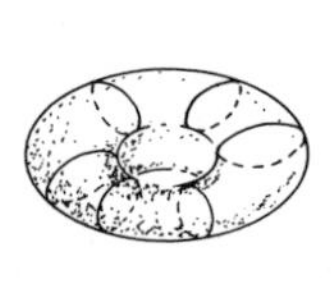

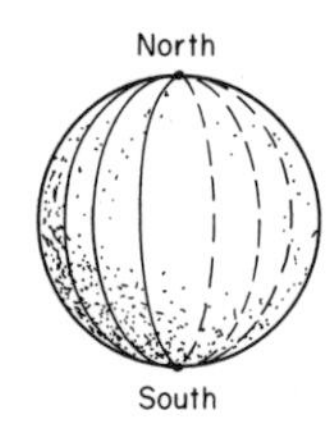

FIGURE 9. *Partition of the torus into smooth, nonintersecting curves. This is impossible for the sphere.*

of exceptional North and South Poles for the geographic lines of latitude and longitude on the surface of the earth.

On the other hand, if a sphere is placed in three-dimensional space, it need have just two horizontal tangent planes (top and bottom). For the torus, this is not the case since "saddle points" must exist "because of the hole in the torus" (Figure 10).

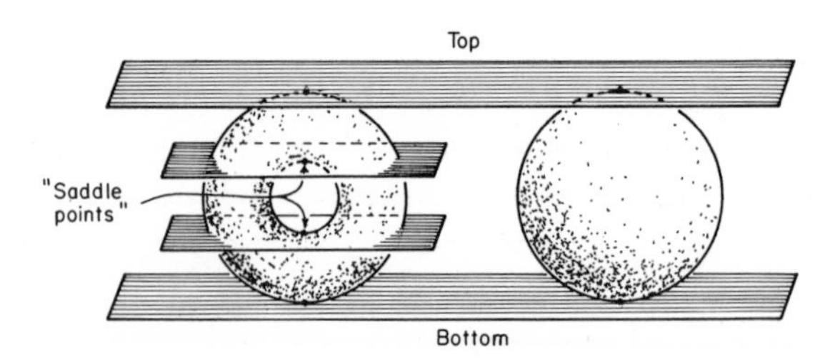

FIGURE 10. *Horizontal tangent planes to the sphere and the torus. The sphere has only two; the torus in general has more.*

Relationships of the sort we have described between the "topology" of surfaces and their functional properties were studied at the end of the nineteenth century by Poincaré in connection with his investigations in celestial mechanics. This line of thought was developed in an interesting new direction by the American mathematician Marston Morse in the

1920's and 30's and has subsequently played an important role in topology.

In the last few years these topological results have been extended to curved infinite-dimensional spaces; this work is, in fact, not yet entirely complete. In the process, it has been necessary to

(i) extend the notion of "smooth curved surface" to infinite-dimensional objects in an appropriate fashion,
(ii) study curves, subsurfaces, etc., in these infinite-dimensional curved surfaces,
(iii) find suitable infinite-dimensional extensions of the finite-dimensional topological principles themselves,
(iv) develop applications of the infinite-dimensional topological theorems obtained; modify and improve the general principles in the light of experience gained with applications of them.

The theory of curved infinite-dimensional spaces, recently developed and currently developing, is often called *nonlinear functional analysis* to stress the fact that it deals with curved spaces and to distinguish it from the older *linear functional analysis*, i.e., from the study of flat infinite-dimensional spaces. We may expect that nonlinear functional analysis will add a new major chapter to mathematics; the future will reveal what the precise contents of this chapter will be.

BIBLIOGRAPHY

Martin Davis, *A First Course in Functional Analysis*, Gordon & Breach, New York, 1967.
V. I. Smirnov, *Linear Algebra and Group Theory*, McGraw-Hill, New York, 1961.
Daniel T. Finkbeiner, *Introduction to Matrices and Linear Transformations*, W. H. Freeman and Co., San Francisco, 1966.

BIOGRAPHICAL NOTE

Jack Schwartz, author of two of the essays in this volume, is a professor at New York University's Courant Institute of Mathematical Sciences. He first became involved in functional analysis during his graduate studies at Yale University and, since then, has written a number of research papers in this field. Although he has been active in the computer field for only a few years, his interest in it actually dates back to his undergraduate days at the City College of New York, when he did some work on computer-related logic.

A great many of the mathematical ideas that apply directly to physics and engineering are collected in the concept of vector spaces. This branch of mathematics has applications in such practical problems as calculating the vibrations of bridges and airplane wings. Logical extensions to spaces of infinitely many dimensions are widely used in modern theoretical physics as well as in many branches of mathematics itself. This essay outlines the concept of vector space and how it started with a few rather obvious notions that continued to grow in mathematical generality and usefulness.

Vector Spaces and Their Applications

E. J. McShane

To the nonmathematician such words as "infinite-dimensional spaces" may seem frighteningly mysterious. How can "space" be infinite-dimensional, or even four-dimensional? There is no reason for mystery; these words refer to rather simple and quite clean-cut ideas. (The reader might also wish to compare the first part of J. T. Schwartz's essay, *Functional Analysis*, which gives a clear presentation of the evolution of these ideas as a process of generalization from the famous theorem of Pythagoras on right triangles.)

Ever since the time of René Descartes, over three hundred years ago, points in a plane have been located by their coordinates with reference to two perpendicular axes intersecting at a point O, called the origin. This is so familiar that we usually speak of "the point $(2, -5)$" instead of "the point whose first coordinate is 2 and whose second coordinate is -5." This is shown in Figure 1. Likewise, we locate points in three-dimensional space by their coordinates with respect to three perpendicular axes, intersecting at a point called the origin.

In general, for a point P in the plane, we can denote its first coordinate by x_1 and its second by x_2. Then P is (x_1, x_2). We also use (x_1, x_2) to symbolize the motion, or *vector*, that carries the origin O to the point P. (The term "vector" is, in fact, derived from the Latin word for "carrier.")

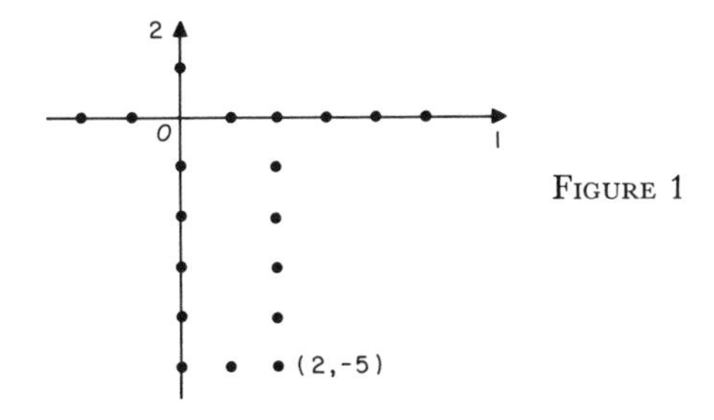

FIGURE 1

We regard as the same vector the motion that carries point A to B the same distance in the same direction, as suggested in Figure 2.

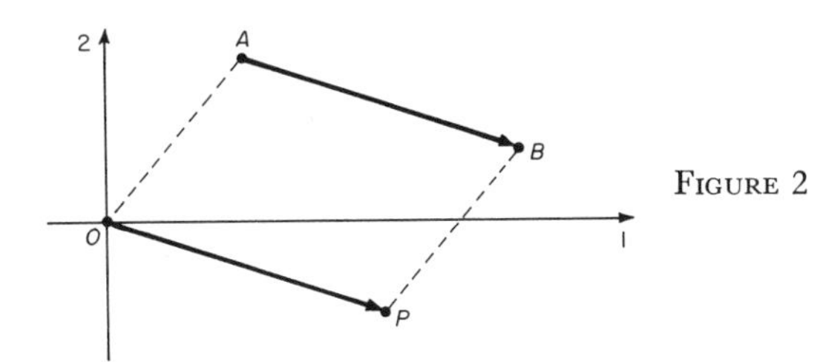

FIGURE 2

In the plane, we shall use the boldface letters **x** and **y** to stand for the vectors (x_1, x_2) and (y_1, y_2), whereas in three-dimensional space **x** and **y** will denote the vectors (x_1, x_2, x_3) and (y_1, y_2, y_3), respectively. If **x** is the vector that carries O to P and **y** is the vector that carries P to Q, it is reasonable to define the "sum" $\mathbf{x} + \mathbf{y}$ of **x** and **y** as the vector that carries O to Q, as shown in Figure 3. (We are interested in final results, not in

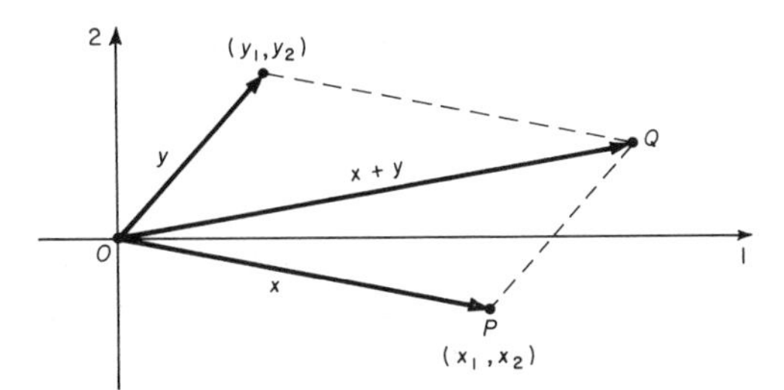

FIGURE 3

"mileage.") From the figure, it is easy to see also that, if $\mathbf{x} = (x_1, x_2)$ and $\mathbf{y} = (y_1, y_2)$, then $\mathbf{x} + \mathbf{y} = (x_1 + y_1, x_2 + y_2)$. If we choose to add **x** to itself, we find that $\mathbf{x} + \mathbf{x} = (2x_1, 2x_2)$, and likewise $\mathbf{x} + \mathbf{x} + \mathbf{x} = (3x_1, 3x_2)$. But it is natural to think of $\mathbf{x} + \mathbf{x}$ as $2\mathbf{x}$ and $\mathbf{x} + \mathbf{x} + \mathbf{x}$ as $3\mathbf{x}$, so that

$$2\mathbf{x} = (2x_1, 2x_2)$$

and

$$3\mathbf{x} = (3x_1, 3x_2).$$

This suggests that for any real number c we should make the definition

$$c\mathbf{x} = (cx_1, cx_2).$$

In three dimensions, we are led by similar suggestions to define $\mathbf{x} + \mathbf{y}$ and $c\mathbf{x}$ by similar formulas, with only the obvious difference that we now need three numbers in the parentheses that denote each vector. In physics, forces and several other important notions are representable by vectors. If a body is acted upon simultaneously by a force $\mathbf{x}$ with components x_1, x_2, x_3 and a force $\mathbf{y}$ with components y_1, y_2, y_3, it will move the same as though acted upon by the single force $\mathbf{x} + \mathbf{y}$ with components $x_1 + y_1$, $x_2 + y_2$, $x_3 + y_3$. Likewise, if a force has components x_1, x_2, x_3, and a second force has the same direction but is c times as strong, that second force has components cx_1, cx_2, cx_3. Thus, our mathematical operations $\mathbf{x} + \mathbf{y}$ and $c\mathbf{x}$ correspond to the physical operations of letting $\mathbf{x}$ and $\mathbf{y}$ act simultaneously and strengthening $\mathbf{x}$ c-fold.

So far, everything we have said, and more, can be summarized in a few short statements. If $\mathbf{x}$, $\mathbf{y}$, and $\mathbf{z}$ are any vectors and a and b are numbers, then

(1) $\mathbf{x} + \mathbf{y}$ and $a\mathbf{x}$ are vectors;
(2) $\mathbf{x} + \mathbf{y} = \mathbf{y} + \mathbf{x}$;
(3) $(\mathbf{x} + \mathbf{y}) + \mathbf{z} = \mathbf{x} + (\mathbf{y} + \mathbf{z})$;
(4) there is a vector $\mathbf{0}$ such that $\mathbf{x} + \mathbf{0} = \mathbf{x}$ for every vector $\mathbf{x}$;
(5) for each vector $\mathbf{x}$ there is a vector $-\mathbf{x}$ such that $\mathbf{x} + (-\mathbf{x}) = \mathbf{0}$;
(6) $a(\mathbf{x} + \mathbf{y}) = a\mathbf{x} + a\mathbf{y}$, $(a + b)\mathbf{x} = a\mathbf{x} + b\mathbf{x}$, $a(b\mathbf{x}) = (ab)\mathbf{x}$;
(7) $1\mathbf{x} = \mathbf{x}$.

Vector Spaces

About the turn of the century, mathematicians noticed that, in the development of both mathematics and its applications, many sets of objects were turning up quite naturally with properties of addition and multiplication in accordance with (1) through (7). Moreover, these properties proved so useful that it was worth while giving a generic name to all collections of objects that possess such addition and multiplication rules. Nowadays they are called *vector spaces* (or *linear spaces*).

The vectors discussed up to this point have been vectors $\mathbf{x}$ with two components x_1, x_2 or with three components x_1, x_2, x_3. Sometimes, however, vectors with a larger number of components arise quite naturally in applications. As a simple example, let us consider a structure (such as a steel bridge) made of elastic members meeting at several junction points, which we choose to label in a definite order. Each junction has a point of rest where it is located when the structure is undisturbed. If we apply a small force to the structure, each junction will move slightly because of the elasticity of the members: The first junction will move x_1 units eastward, x_2 units northward, x_3 units upward; the second junction will move x_4 units eastward, x_5 units northward, x_6 units upward; and so on. These numbers may be negative. If, for instance, the second junction moves a tenth of a unit westward, then $x_4 = -0.1$. To express the whole deformation of the structure, we need three times as many numbers as there are junctions, because each junction point has three coordinates. We can denote these by $x_1, x_2, \cdots, x_n$, where n is three times the number of junctions; and if we keep in mind what n is, we can abbreviate this to $\mathbf{x}$:

$$\mathbf{x} = (x_1, x_2, \cdots, x_n).$$

These symbols can thus be used to specify the deformations of the structure from its rest position.

Now, by analogy with the case of vectors with two or three components, we can define the sum of vectors $\mathbf{x} = (x_1, x_2, \cdots, x_n)$ and $\mathbf{y} = (y_1, y_2, \cdots, y_n)$ to be

$$\mathbf{x} + \mathbf{y} = (x_1 + y_1, x_2 + y_2, \cdots, x_n + y_n),$$

and for each real number c we can define

$$c\mathbf{x} = (cx_1, cx_2, \cdots, cx_n).$$

Statements (1) through (7) are easy to verify, and so the collection of all such "n-tuples" of numbers $(x_1, \cdots, x_n)$ forms a vector space. It turns out that there is a good reason — a little deeper than the obvious fact that each vector $\mathbf{x} = (x_1, \cdots, x_n)$ has n components — for labeling this vector space *n-dimensional*. (Very briefly, a vector space is called n-dimensional if in it there are n vectors having the property that every vector in the space is a sum of multiples of these n vectors, but no smaller number of vectors than n has this property.)

Oscillations of an Elastic Structure

Up to this point, we have indicated how the deformations of an elastic structure, such as a steel bridge or an airplane wing, can be described by vectors $\mathbf{x} = (x_1, \cdots, x_n)$ in a vector space of an appropriate number n of dimensions. We shall now outline how the vector-space point of view simplifies the analysis of the oscillations or vibrations of such a structure.

We start with a simple case as a guide. Suppose that a straight steel rod of irregular cross section is clamped at one end, say horizontally. If we pull the free end away from its rest position, we store energy in the rod.

If we pull it k times as far, we not only have to produce k times the displacement, we also have to apply k times the force, because this force is proportional to the displacement. Thus, the energy required is k^2 times as great and hence is a quadratic function of the amount of displacement. This makes it plausible (and by physics it is indeed true) that, if we displace the end of the rod x_1 units horizontally and x_2 units vertically, the energy E stored in the rod will be a quadratic function of x_1 and x_2; that is,

$$E = ax_1^2 + 2bx_1x_2 + cx_2^2,$$

where a, b, and c are real numbers. Again by physics, if we release the displaced end, it will start to move "downhill," that is, in the direction in which E decreases most rapidly. It turns out that, in general, this will not be straight toward the origin, and the end of the rod will, in general, describe a complicated curve called a "Lissajous figure."

In a word, the simplification we are going to describe depends on a strategically selected change of coordinate system in the vector space concerned. For our example of a clamped vibrating rod, this vector space is a plane perpendicular to the rod in its horizontal rest position. At any given instant, the position of the free end of the rod is specified by a vector $\mathbf{x} = (x_1, x_2)$ in this plane; the vector's components x_1 and x_2 give the horizontal and vertical displacements from the rest position, as suggested in Figure 4.

To describe our simplification, we need to define the ideas of *length* and *perpendicularity* of vectors. For vectors in a plane, these ideas have a familiar intuitive meaning. In fact, Figure 4 suggests how the length of the

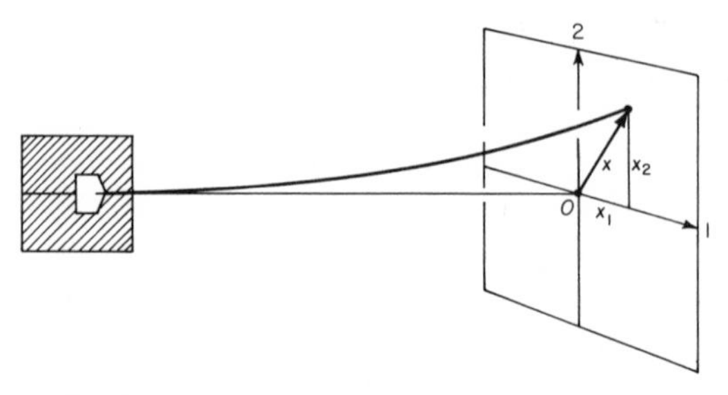

FIGURE 4

vector $\mathbf{x} = (x_1, x_2)$ should be defined: in accordance with the theorem of Pythagoras, this length, $||x||$, squared should be given by

$$||\mathbf{x}||^2 = x_1^2 + x_2^2$$

so that

$$||\mathbf{x}|| = \sqrt{x_1^2 + x_2^2}.$$

The essay of J. T. Schwartz, referred to at the outset, indicates how we arrive at natural generalizations of this formula: for a vector $\mathbf{x} = (x_1, x_2, x_3)$ with three components, we find that the length (or *norm*, as it is sometimes called) is given by

$$||x|| = \sqrt{x_1^2 + x_2^2 + x_3^2},$$

while for a vector $\mathbf{x} = (x_1, x_2, \cdots, x_n)$ with n components, we have

$$||\mathbf{x}|| = \sqrt{x_1^2 + x_2^2 + \cdots + x_n^2}.$$

Schwartz's essay also shows (using slightly different notation) how it happens that two vectors $\mathbf{x} = (x_1, x_2)$ and $\mathbf{y} = (y_1, y_2)$ turn out to be perpendicular (or *orthogonal*) if and only if

$$x_1y_1 + x_2y_2 = 0.$$

For vectors $\mathbf{x} = (x_1, x_2, x_3)$ and $\mathbf{y} = (y_1, y_2, y_3)$, the natural generalization of this definition of orthogonality turns out to be

$$x_1y_1 + x_2y_2 + x_3y_3 = 0$$

and, for vectors $\mathbf{x} = (x_1, x_2, \cdots, x_n)$ and $\mathbf{y} = (y_1, y_2, \cdots, y_n)$, this definition becomes

$$x_1y_1 + x_2y_2 + \cdots + x_ny_n = 0.$$

Now we return to our example of the clamped rod. All of our analysis holds just as well for one set of perpendicular coordinate axes as for another. Thus if $\mathbf{u}_1$ and $\mathbf{u}_2$ are each vectors of length 1 and are perpendicular, we can express each vector $\mathbf{x}$ as some multiple X_1 of $\mathbf{u}_1$ plus some multiple X_2 of $\mathbf{u}_2$:

$$\mathbf{x} = X_1\mathbf{u}_1 + X_2\mathbf{u}_2,$$

and we may think of X_1 and X_2 as components of $\mathbf{x}$ with respect to coordinate axes in the directions of $\mathbf{u}_1$ and $\mathbf{u}_2$. The value of E at a given place is unchanged, but the formula that expresses it in terms of X_1 and X_2 will look different from the formula in terms of x_1 and x_2. Ever since the time of Descartes, it has been known that it is possible to choose the perpendicular vectors $\mathbf{u}_1$, $\mathbf{u}_2$ of length 1 in such a way that the expression for E simplifies to

$$E = AX_1^2 + CX_2^2,$$

where A and C are real numbers that must be positive because every nonzero bending of the rod requires positive energy. The lines through the rest point O in directions $\mathbf{u}_1$ and $\mathbf{u}_2$ are called the *principal axes*.

It now turns out that, if we pull the end of the rod to a point $X_1\mathbf{u}_1$ on the first principal axis and then release it, the direction in which E decreases most rapidly is *straight toward* O. Thus, the end of the rod will stay on this axis and never be pulled to either side; its motion will be a simple pendulum motion back and forth along the axis. Also, if we know the formula for the frequency of a pendulum (vibrations per unit time), we can figure that the frequency of the vibration is a calculable number times $\sqrt{A}$. In the same way, from a start along the second principal axis, the end of the rod will make pendulumlike oscillations along that principal axis with frequency proportional to $\sqrt{C}$.

By applying a similar argument to a more complicated elastic structure, whose deformations are given by a vector $\mathbf{x} = (x_1, x_2, \cdots, x_n)$, we find again by physics that the energy E is (with negligible error) a quadratic function of $x_1, \cdots, x_n$. That is, E is a sum of terms each of which is a multiple of the product of one of $x_1, \cdots, x_n$ and itself or another of $x_1, \cdots, x_n$. Next, we would very much like to find such vectors $\mathbf{u}_1, \cdots, \mathbf{u}_n$ that have the property that, when we express $\mathbf{x}$ in the form

$$\mathbf{x} = X_1\mathbf{u}_1 + X_2\mathbf{u}_2 + \cdots + X_n\mathbf{u}_n,$$

the energy formula simplifies to

$$E = A_1X_1^2 + A_2X_2^2 + \cdots + A_nX_n^2,$$

in analogy with the two-dimensional case. Here we draw on a result in pure mathematics: The "principal axis theorem" says it is always possible to find such vectors. Now we are as well off as we were in the simple case. If we give the structure a displacement $\mathbf{x}$ that can be represented as $X_1\mathbf{u}_1$, the system will start moving in the direction in which the energy diminishes most rapidly, straight toward O. The oscillation will be back and forth along the first principal axis, with frequency proportional to $\sqrt{A_1}$. The same is true for the motions starting on any of the other principal axes; and all motions of the system are compounded of these simple motions.

The "natural frequencies," which are proportional to $\sqrt{A_1}$, $\sqrt{A_2}$, and so on, are very important. When the system is subjected to a rhythmic disturbance whose frequency matches one of these natural frequencies, the size of the oscillations can build up dangerously. As a grim example, we have the wings of the first Electra airplanes.

Perpendicular Vectors

We have already noted that two n-dimensional vectors $\mathbf{x}$ and $\mathbf{y}$ are perpendicular (or orthogonal, as we often say) if and only if a certain combination

$$x_1y_1 + x_2y_2 + \cdots + x_ny_n$$

of their components is zero. For brevity let us use the notation $\langle \mathbf{x}, \mathbf{y} \rangle$ for $x_1y_1 + x_2y_2$ in the two-dimensional case, for $x_1y_1 + x_2y_2 + x_3y_3$ in the three-dimensional case, and for $x_1y_1 + x_2y_2 + \cdots + x_ny_n$ in the n-dimensional case. Then it is easy to see that, whenever $\mathbf{x}$, $\mathbf{y}$, and $\mathbf{z}$ are vectors and a and b are numbers, the following three properties hold:

$$\langle a\mathbf{x} + b\mathbf{y}, z \rangle = a \langle \mathbf{x}, \mathbf{z} \rangle + b \langle \mathbf{y}, \mathbf{z} \rangle; \tag{P1}$$

$$\langle \mathbf{y}, \mathbf{x} \rangle = \langle \mathbf{x}, \mathbf{y} \rangle; \tag{P2}$$

$$\text{if } \mathbf{x} \text{ is not } \mathbf{0}, \text{ then } \langle \mathbf{x}, \mathbf{x} \rangle \text{ is positive.} \tag{P3}$$

In this notation, vectors $\mathbf{x}$ and $\mathbf{y}$ are orthogonal (perpendicular) if and only if

$$\langle \mathbf{x}, \mathbf{y} \rangle = 0.$$

Lengths, too, are neatly expressed in this notation. For a vector $\mathbf{x} = (x_1, x_2)$ in the plane, we have already indicated that its length, or *norm*, is given by

$$||\mathbf{x}|| = \sqrt{x_1^2 + x_2^2};$$

since $\langle \mathbf{x}, \mathbf{x} \rangle = x_1^2 + x_2^2$, we obtain the simple formula

$$||\mathbf{x}|| = \sqrt{\langle \mathbf{x}, \mathbf{x} \rangle}.$$

The same formula is easily seen to hold for vectors $\mathbf{x} = (x_1, x_2, x_3)$ in three-dimensional space and for vectors $\mathbf{x} = (x_1, x_2, \cdots, x_n)$ in n-dimensional space.

In general, vector spaces in which there is assigned to each pair $\mathbf{x}$, $\mathbf{y}$ of vectors a number $\langle \mathbf{x}, \mathbf{y} \rangle$ in such a way that the above three properties hold are called *inner product spaces*, and the number $\langle \mathbf{x}, \mathbf{y} \rangle$ is called the *inner product* of $\mathbf{x}$ and $\mathbf{y}$. In the next section, we shall try to give a hint of the great variety of useful inner product spaces. Some of these are infinite-dimensional, and many spring from ideas entirely outside the "space" context. It is here that the introduction of geometric terms such as norm (length) and orthogonality (perpendicularity) proves especially profitable. In *any* inner product, we *define* orthogonality of vectors $\mathbf{x}$ and $\mathbf{y}$ by the condition

$$\langle \mathbf{x}, \mathbf{y} \rangle = 0,$$

and we *define* the norm $||\mathbf{x}||$of a vector $\mathbf{x}$ by setting

$$||x|| = \sqrt{\langle \mathbf{x}, \mathbf{x} \rangle}.$$

In the plane and three-dimensional space, this agrees with our earlier ideas and, in other cases, this geometric language helps the minds of mathematicians. When we speak of length or perpendicularity in any inner product space, we are automatically reminded of the simplest cases, namely the plane and three-dimensional space, in which we are familiar with a multitude of theorems involving these concepts. We naturally try to see whether similar theorems hold in the more complicated spaces too.

Infinite-Dimensional Spaces

What are some of these more complicated spaces? J. T. Schwartz's essay, referred to at the beginning, indicates how we may pass from vectors

$$\mathbf{x} = (x_1, x_2, \cdots, x_n)$$

with n components to vectors

$$\mathbf{x} = (x_1, x_2, \cdots)$$

with infinitely many components. With the natural definitions

$$\mathbf{x} + \mathbf{y} = (x_1 + y_1, x_2 + y_2, \cdots)$$

and

$$c\mathbf{x} = (cx_1, cx_2, \cdots),$$

these form a vector space. This vector space turns out to be of no finite number of dimensions and hence is called infinite-dimensional.

A vector $\mathbf{x} = (x_1, x_2, \cdots)$ has finite length given by the natural formula

$$||\mathbf{x}|| = \sqrt{x_1^2 + x_2^2 + \cdots}$$

only when the sum of the infinitely many terms $x_1^2 + x_2^2 + \cdots$ is finite. For vectors $\mathbf{x} = (x_1, x_2, \cdots)$ and $\mathbf{y} = (y_1, y_2, \cdots)$ of finite length, it turns out that the sum

$$\langle \mathbf{x}, \mathbf{y} \rangle = x_1 y_1 + x_2 y_2 + \cdots$$

is also finite and defines a properly behaved inner product, so that all such vectors taken together form an inner product space. This particular inner product space is called a *Hilbert space*, after David Hilbert, who was widely regarded as the most distinguished mathematician of his generation and who pioneered the systematic study of such spaces early in the twentieth century.

About 100 years earlier, an applied mathematician, Joseph Fourier, had made a path-breaking study of certain differential equations governing the conduction of heat. He was able to find solutions of such equations in the form of infinite series of trigonometric sine and cosine functions.

In the succeeding years of the nineteenth century, there evolved a whole theory of Fourier series, as they came to be called; and comparable series in which sines and cosines were replaced by other special functions (including those associated with the names Legendre, Jacobi, Laguerre, Hermite, Bessel, and others) were found to yield solutions of many differential-equation problems arising in physics and engineering. All these developments were clarified, unified, and extended by the vector-space point of view.

To get a better idea of this, consider a function such as the one graphed in Figure 5, which assigns to each number t between 0 and 1 a functional

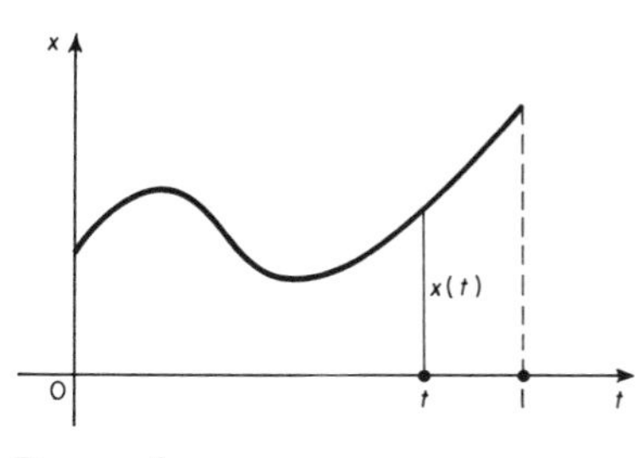

FIGURE 5

value $x(t)$. We regard this function as a vector $\mathbf{x}$ with infinitely many components $x(t)$, one for each number t between 0 and 1. With natural definitions for the sum $\mathbf{x} + \mathbf{y}$ of two such functions and for the product $c\mathbf{x}$ of such a function with a number c, all such functions taken together form a vector space. Furthermore, if we restrict attention to sufficiently well-behaved functions, these form the vectors of an inner product space. As in other cases we have discussed, the inner product $\langle \mathbf{x}, \mathbf{y} \rangle$ is obtained by forming the products $x(t)y(t)$ of corresponding components of the vectors $\mathbf{x}$ and $\mathbf{y}$ and summing, except that now the summing is done by the process of integration:

$$\langle \mathbf{x}, \mathbf{y} \rangle = \int_0^1 x(t)y(t)\, dt.$$

As before, orthogonality of vectors $\mathbf{x}$ and $\mathbf{y}$ is defined by the condition $\langle \mathbf{x}, \mathbf{y} \rangle = 0$; it is in this connection that we may begin to grasp the vector space unification of the theories of Fourier series and related developments. The first fact to be noted is that the (infinitely many) sine and cosine functions of Fourier series, thought of as vectors in an infinite-dimensional inner product space, all turn out to be orthogonal to one another. Now, suppose we are given the differential equation that governs, say, the flow of heat along a uniform insulated metal rod, where the temperature at some initial instant of time is known in advance for each point along the rod. The problem is to predict the temperature at later instants at each point along the rod. For each such later instant, the solution function emerges as the sum of an infinite series of multiples of Fourier's sine and cosine functions. From the vector-space point of view, this solution function is a vector; and these multiples are the components of this vector with respect to "coordinate axes" in the mutually orthogonal directions defined by these sine and cosine functions, when they, too, are thought of as vectors.

Now, just as in the case of the sines and cosines of Fourier, the classes of special functions of Legendre, Jacobi, Laguerre, Hermite, Bessel, and so on, may each be regarded as defining a coordinate system in an appropriate infinite-dimensional inner product space. Finding the solution of a differential-equation problem of physics in terms of infinite series of such functions may be regarded as finding the components of the solution vector with respect to this coordinate system.

To conclude this section, we mention briefly an extremely important mathematical development which these same infinite series of functions helped to stimulate. This was the invention, by Henri Lebesgue early in the twentieth century, of a more powerful kind of integral. Since inner products in function spaces are defined by integrals, Fourier series and other such infinite series tend to involve integrals. In such series it was found that the earlier notion of integral, which had been clarified in the mid-nineteenth century by the great mathematician Riemann, possessed inadequate convergence properties and applied to too limited a class of functions. Both these defects were remedied in the Lebesgue integral, which subsequently proved to be of prime importance in probability theory and many other parts of mathematics.

Vector Spaces and Quantum Mechanics

The two major developments of twentieth-century physics have been relativity theory and quantum mechanics. The second has come to involve infinite-dimensional vector spaces in an essential way, but it is hardly possible to describe this in any detail in a few pages. For the most part, therefore, we shall content ourselves with statements in very general terms.

Quantum mechanics dates back to independent work of the theoretical physicists Erwin Schrödinger and Werner Heisenberg in the early 1920's. Schrödinger's work revolved about a certain differential equation, while Heisenberg's was phrased in terms of matrices. Despite these and other apparent differences, the two theories were soon recognized as essentially equivalent, and in 1927 a definitive mathematical treatment was given by the mathematician John Von Neumann. The appropriate mathematical setting turned out to be the theory of linear operators in a (complex) Hilbert space.

It is easy to say what linear operators are. Every high school student has had to solve equations

$$ax_1 + bx_2 = g_1 \qquad \text{and} \qquad cx_1 + dx_2 = g_2$$

for the "unknowns" x_1, x_2 in terms of g_1 and g_2, given the coefficients a, b, c, d. We reword this: Where a, b, c and d are given, corresponding to each vector $\mathbf{x} = (x_1, x_2)$ we can construct a new vector whose first component is $ax_1 + bx_2$ and whose second component is $cx_1 + dx_2$. For this new vector we use the notation $T\mathbf{x}$ since it is a "transform" of $\mathbf{x}$. The problem now is: Given a vector $\mathbf{g} = (g_1, g_2)$, find $\mathbf{x}$ so that $T\mathbf{x} = \mathbf{g}$. The problem of solving n equations for n unknowns can be written in this same compact form, $\mathbf{x}$ and $T\mathbf{x}$ now belonging to the n-dimensional space of vectors $(x_1, x_2, \cdots, x_n)$.

The transformation T is easily seen to have the properties

$$T(\mathbf{x} + \mathbf{y}) = T\mathbf{x} + T\mathbf{y} \qquad \text{and} \qquad T(a\mathbf{x}) = a(T\mathbf{x})$$

for all vectors $\mathbf{x}$ and $\mathbf{y}$ and all real numbers a. It is those transformations T, defined in a vector space and having these two properties, that are called *linear operators*. There are many examples of linear transformations, or operators, in function spaces. For instance, a phonograph amplifier transforms an input function, the voltage across the pick-up head, into an output function, the voltage across the speaker terminals. The transformation is, at least approximately, linear; if the output function is markedly different in shape from the input one, the amplifier is "low fidelity."

Let us try to give at least some indication of how these linear operators are used in quantum theory. For these we need to consider a certain infinite-dimensional function space, which is in fact a Hilbert space of the sort mentioned earlier but with this difference. Now both the values of functions that are the vectors of the space and the numbers by which we are permitted to multiply these vectors are *complex* numbers, numbers of the form $a + b\sqrt{-1}$. (For complex inner-product spaces, we also need to modify our definition of inner product slightly.)

Next, suppose we are experimenting with a system that might be an atom or a molecule or a crystal. For each state the system can be in, there is a corresponding vector ψ in this Hilbert space. Furthermore, for each experimental procedure, there is a corresponding linear operator T. Further, if the experiment with operator T is performed on the system when it is in state ψ, the expected value of the result of the experiment is the inner product $\langle T\psi, \psi \rangle$. As in the study of the vibrating structures discussed earlier, the vectors that transform into multiples of themselves are particularly important. Physically, these correspond to experimental results that are unchanging in time. That is, they describe so-called "steady states." A system cannot go from one steady state with energy E_0 to another with energy E_1 unless it either emits or absorbs energy. With atoms and molecules, this emitted or absorbed energy shows up as the emission or absorption lines in the spectrum. The spectrum of the hydrogen atom has very many lines, and it is most gratifying that, by proper choice of a certain parameter in the state vector, we can account for all the lines and for other experimental results, too.

This mathematical theory can be refined to take account of minute effects like electron spin and relativistic corrections to give extremely accurate results. It can also be extended to complicated atoms and to quantized fields; but here much remains to be done.

It is not easy to predict the future uses of infinite-dimensional spaces. Since the fundamental idea is to extend geometric intuitions to new problems, the application may occur whenever someone has a good intuition. Recently, the mathematics of computation has been strongly flavored with such ideas. All that is safe to say is that these uses are likely to appear almost anywhere.

BIBLIOGRAPHY

P. R. Halmos, *Finite-Dimensional Vector Spaces*, Van Nostrand, Princeton, N.J., 1958.

R. V. Churchill, *Fourier Series and Boundary Value Problems*, McGraw-Hill, New York, 1963.

(Also see J. T. Schwartz's essay, *Functional Analysis*, and its bibliography.)

BIOGRAPHICAL NOTE

E. J. McShane was born in New Orleans, Louisiana. Following graduate education at the University of Chicago and short periods at Göttingen and Princeton, he has spent the major part of his career as Professor of Mathematics at the University of Virginia. He has always had an interest in applications, and his first college degrees were in engineering and physics. His research has been primarily in calculus of variations and integration theory, his most recent contributions being concerned with probability integrals and applications of calculus of variations in control theory. During World War II his distinguished work in exterior ballistics at the Army's Aberdeen Proving Ground exploited the vector-space point of view.

Some of the most abstruse concepts of mathematics have an uncanny way of becoming essential tools in physics. Many physicists have been so impressed by the usefulness of mathematics that they have attributed to it almost mystical power. Kepler said that all nature is symbolized in the art of geometry. Hertz felt that mathematical formulas had an intelligence of their own. And Einstein built the theory of general relativity on a mathematical leap into the dark. Today physicists rely on mathematics not merely as a tool for calculation but as a source of inspiration. The author, a noted theoretical physicist, describes three mathematical approaches that are helping to clear up the confusion surrounding the elementary particles and to explain the grand scheme into which they all must fit.

Mathematics in the Physical Sciences

Freeman J. Dyson

In 1910 the mathematician Oswald Veblen and the physicist James Jeans were discussing the reform of the mathematical curriculum at Princeton University. "We may as well cut out group theory," said Jeans. "That is a subject which will never be of any use in physics." It is not recorded whether Veblen disputed Jeans's point, or whether he argued for the retention of group theory on purely mathematical grounds. All we know is that group theory continued to be taught. And Veblen's disregard for Jeans's advice turned out to be of some importance to the history of science at Princeton. By an irony of fate group theory later grew into one of the central themes of physics, and it now dominates the thinking of all of us who are struggling to understand the fundamental particles of nature. It also happened by chance that Hermann Weyl and Eugene P. Wigner, who pioneered the group-theoretical point of view in physics from the 1920's to the present, were both Princeton professors.

This little story has several morals. The first moral is that scientists ought not to make off-the-cuff pronouncements concerning matters outside their special field of competence. Jeans provides us with a clear lesson on the evil effects of the habit of pontification. Starting from this unfortunate beginning with Veblen, he later went from bad to worse, becoming a successful popular writer and radio broadcaster, accepting a knighthood

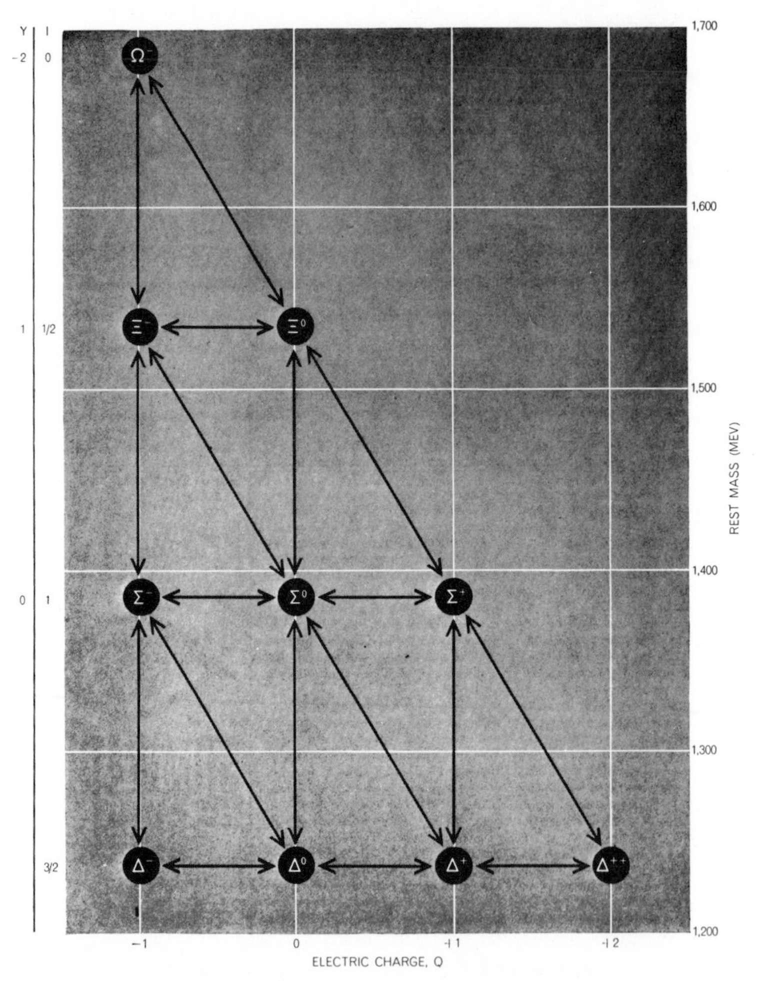

FIGURE 1. *Success of group theory in the physics of fundamental particles was dramatized early in 1964 with the discovery of the omega-minus* (Ω^-) *baryon at the Brookhaven National Laboratory. The existence of the omega minus had been predicted by the "eightfold way," a theory devised independently by Murray Gell-Mann and Yuval Ne'eman. The term "eightfold" refers to a classification scheme based on the mathematical theory of abstract groups. Previous theory had shown that "isotopic spin" symmetry* (horizontal arrows) *connects families of particles with different values of electric charge* (Q). *The eightfold way invokes a new system of symmetries* (vertical and slanted arrows) *to group together superfamilies of particles with different values of hypercharge* (Y) *and isotopic spin* (I). *The omega-minus baryon was needed to complete a superfamily of 10 members, of which nine members were previously known: a delta* (Δ) *quartet, a sigma* (Σ) *triplet and a xi* (Ξ) *doublet. The omega minus is the only baryon singlet with a negative electric charge, and its observed mass is within a few million electron volts (mev) of the mass predicted by theory.*

and ruining his professional reputation with suave and shallow speculations on religion and philosophy.

We ought not, however, to look so complacently on the decline and fall of Jeans. There, but for the grace of God, go we. After all, Jeans in 1910 was a respected physicist (although Princeton, aping the English custom in titles as in pseudo-Gothic architecture, called him professor of applied mathematics). He was neither more incompetent nor more ignorant than most of his colleagues. Very few men at that time had the slightest inkling of the fruitfulness that would result from the marriage of physics and group theory. So the second and more serious moral of our story is that the future of science is unpredictable. The place of mathematics in the physical sciences is not something that can be defined once and for all. The interrelations of mathematics with science are as rich and various as the texture of science itself.

One factor that has remained constant through all the twists and turns of the history of physical science is the decisive importance of mathematical imagination. Each century had its own particular preoccupations in science and its own particular style in mathematics. But in every century in which major advances were achieved the growth in physical understanding was guided by a combination of empirical observation with purely mathematical intuition. For a physicist mathematics is not just a tool by means of which phenomena can be calculated; it is the main source of concepts and principles by means of which new theories can be created.

All through the centuries the power of mathematics to mirror the behavior of the physical universe has been a source of wonder to physicists. The great seventeenth-century astronomer Johannes Kepler, discoverer of the laws of motion of the planets, expressed his wonder in theological terms: "Thus God himself was too kind to remain idle, and began to play the game of signatures, signing his likeness into the world; therefore I chance to think that all nature and the graceful sky are symbolized in the art of geometry." In the more idealistic nineteenth century the German physicist Heinrich Hertz, who first verified James Clerk Maxwell's electromagnetic equations by demonstrating the existence of radio waves, wrote: "One cannot escape the feeling that these mathematical formulae have an independent existence and an intelligence of their own, that they are wiser than we are, wiser even than their discoverers, that we get more out of them than was originally put into them." Lastly, in our rationalistic twentieth century Eugene Wigner has expressed his puzzlement at the success of more modern mathematical ideas in his characteristically dry and modest manner: "We are in a position similar to that of a man who was provided with a bunch of keys and who, having to open several doors in succession, always hit on the right key on the first or second trial. He became skeptical concerning the uniqueness of the coordination between keys and doors."

The mathematics of Kepler, the mathematics of Hertz and of Wigner have almost nothing in common. Kepler was concerned with Euclidean geometry, circles and spheres and regular polyhedra. Hertz was thinking of partial differential equations. Wigner was writing about the use of complex numbers in quantum mechanics, and no doubt he was also

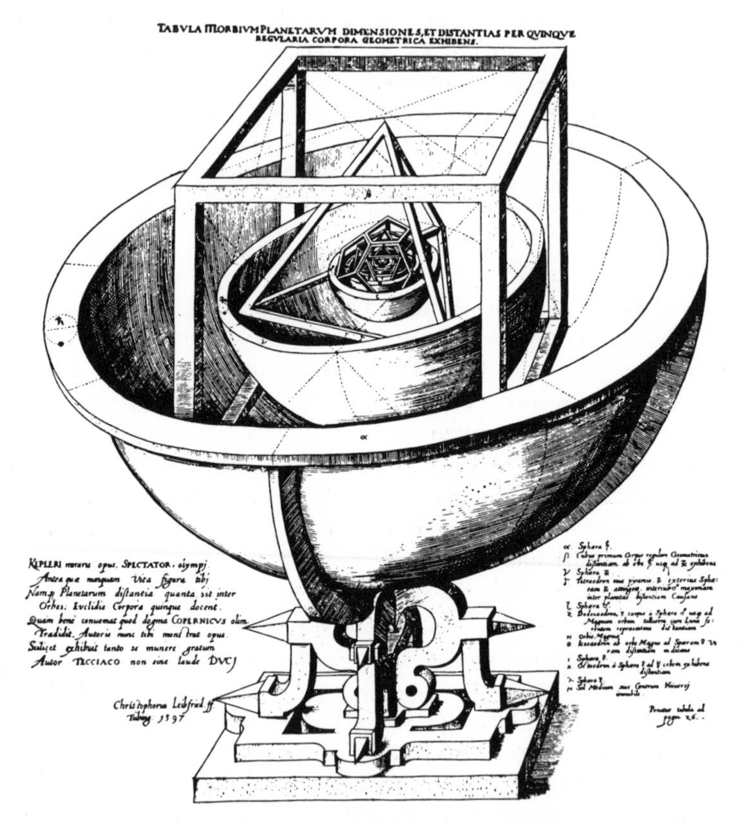

FIGURE 2. *Kepler's model of the solar system, published in 1596, was based on the five "perfect" solids of Euclidean geometry. The planetary orbits were successively inscribed in and circumscribed about an octahedron, an icosahedron, a dodecahedron, a tetrahedron, and a cube. The model is a supreme example of misguided mathematical intuition. Although Kepler was aware of the discrepancies between his theory and the best observations of his time, he always regarded this model as one of his greatest achievements.*

thinking about (but not mentioning) his own triumphant introduction of group theory into many diverse areas of physics. Euclid, partial differential equations and group theory are three branches of mathematics so remote from each other that they seem to belong to different mathematical universes. And yet all three of them turn out to be intimately involved in our one physical universe. These are astonishing facts, understood fully by nobody. Only one conclusion seems to follow with assurance from such facts. The human mind is not yet close to any complete under-

RATIO OF ORBITS	COPERNICAN VALUES	KEPLER'S MODEL	MODERN VALUES
MERCURY MAXIMUM / VENUS MINIMUM	.723	.707	.650
VENUS MAXIMUM / EARTH MINIMUM	.794	.795	.741
EARTH MAXIMUM / MARS MINIMUM	.757	.795	.735
MARS MAXIMUM / JUPITER MINIMUM	.333	.333	.337
JUPITER MAXIMUM / SATURN MINIMUM	.635	.577	.604

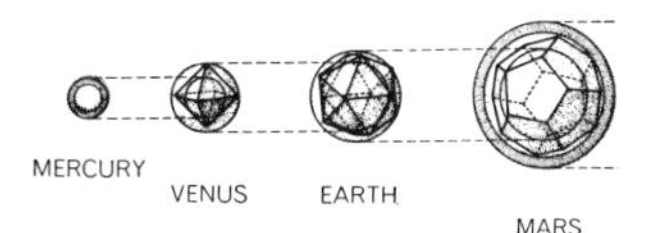

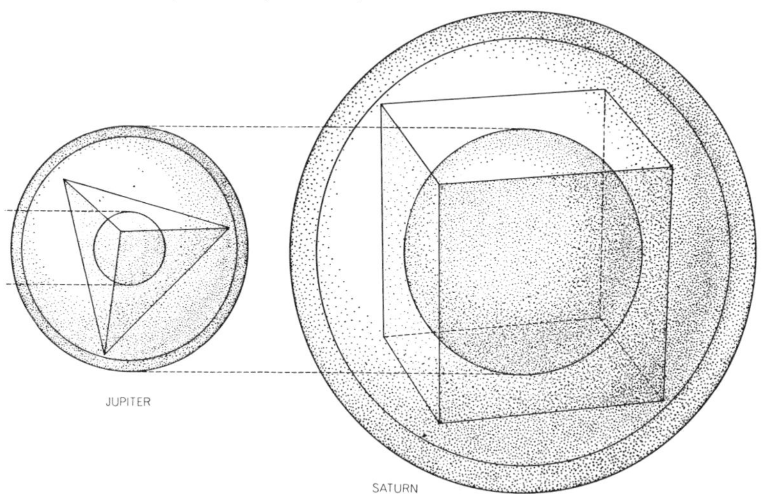

FIGURE 3. *Exploded view of Kepler's polyhedron model of the solar system shows how each planetary orbit was supposed to occupy a spherical shell whose thickness corresponded to the difference between that planet's maximum and minimum distance from the sun. The table contains three sets of values for the ratio between each planet's maximum orbit and the next outer planet's minimum orbit. The first column gives the observational values obtained by Kepler from Copernicus. The second column gives the theoretical values predicted by Kepler's polyhedron model. The third column gives the accepted modern values. Kepler cheated in the case of Mercury in order to account for the most conspicuous discrepancy between his theory and the Copernican values: although the four outer polyhedra are circumscribed around a planetary shell in the usual way (the shell touches the faces of the polyhedron), the octahedron is circumscribed around the shell of Mercury in a special way (the shell touches the edges of the octahedron).*

standing of the physical world, or of the mathematical world, or of the relations between them.

In this article I shall not attempt any deep philosophical discussion of the reasons why mathematics supplies so much power to physics. In each century it is only a few physicists — in our century perhaps only Albert Einstein, Weyl, Niels Bohr, P. W. Bridgman, and Wigner — who dig deep enough into the foundations of our knowledge to reach genuinely philosophical difficulties. The vast majority of working scientists, myself

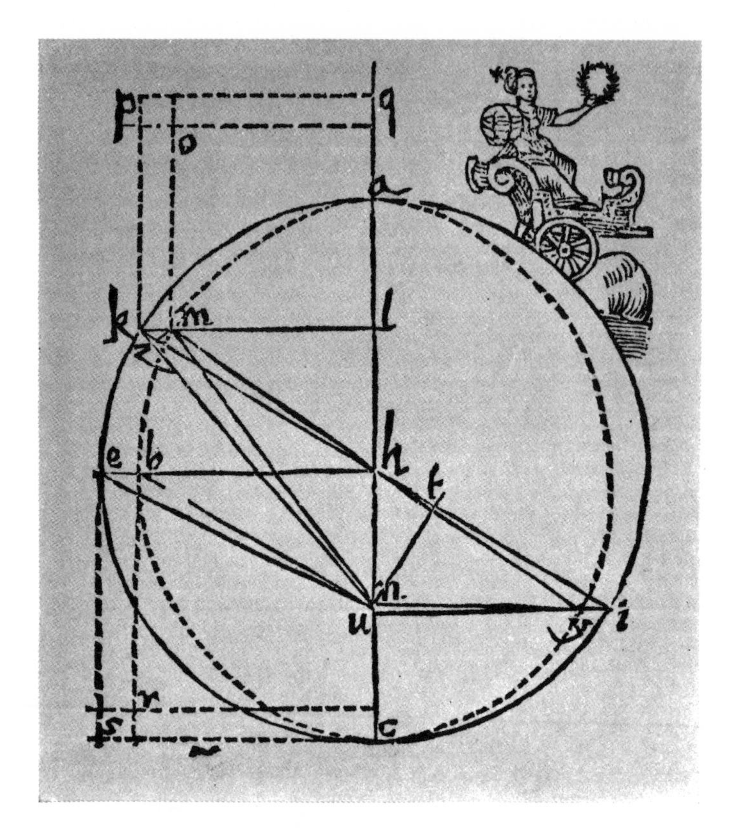

FIGURE 4. *Discovery of elliptical orbit for Mars was Kepler's great triumph after years of trying to make circular orbits satisfy Tycho Brahe's observations. In this diagram he shows that Mars sweeps out equal areas, measured from the sun at* n, *in equal times.*

included, find comfort in the words of the French mathematician Henri Lebesgue: "In my opinion a mathematician, in so far as he is a mathematician, need not preoccupy himself with philosophy — an opinion, moreover, which has been expressed by many philosophers."

We are content to leave the philosophizing to giants such as Bohr and Wigner, while we amuse ourselves with the exploration of nature on a more superficial level. I shall accordingly not discuss further the ultimate reasons why mathematical concepts have come to be preeminent in physics. I shall beg the philosophical question, assuming as an article of faith that nature is to be understood in mathematical terms. The questions I shall address are practical ones relating to the way in which mathematical ideas react on physics. What are the standards of taste and judgment that mathematics imposes on the physicist? Which are the parts of mathematics that now offer hope for new physical understanding? In conclusion, since one concrete example is better than a mountain of prose,

I shall sketch the role group theory has played in physics, leading up to the theory of fundamental particles known as the "eightfold way" [see "Strongly Interacting Particles," by Geoffrey F. Chew, Murray Gell-Mann, and Arthur H. Rosenfeld; *Scientific American*, February 1964]. This theory, developed independently by Gell-Mann and Yuval Ne'eman, has been brilliantly vindicated by the discovery of the omega-minus particle.

Before plunging into the details of present-day problems, I shall illustrate the effects of mathematical tastes and prejudices on physics with some historical examples. In trying to explain technical matters to a nontechnical audience, it is often helpful to examine past history and draw analogies between the problems of the past and those of the present. The reader should be warned not to take historical analogies too seriously. Very few active scientists are particularly well informed about the history of science, and almost none are directly guided in their work by historical analogies. In this respect scientists can be compared to politicians. The greatest politician of our century was probably Lenin, and he operated successfully within a historical viewpoint that was grossly limited and distorted. The only important historian of modern times to achieve high political office was François Guizot, prime minister of France during the 1840's, and all his historical understanding did not save him from mediocrity as a statesman. A good historian is too much committed to the past to be either a creative political leader or a creative scientist. In science at least, if a man wishes to achieve greatness, he should follow the advice of William Blake: "Drive your cart and your plow over the bones of the dead."

The most spectacular example in physics of the successful use of mathematical imagination is still Einstein's theory of gravitation, otherwise known as the general theory of relativity. To build his theory Einstein used as his working material non-Euclidean geometry, a theory of curved spaces that had been invented during the nineteenth century. Einstein took the revolutionary step of identifying our physical space-time with a curved non-Euclidean space, so that the laws of physics became propositions in a geometry radically different from the classical flat-space geometry. All this was done by Einstein on the basis of very general arguments and aesthetic judgments. The observational tests of the theory were made only after it was essentially complete, and they did not play any part in the creative process. Einstein himself seems to have trusted his mathematical intuition so firmly that he had no feeling of nervousness about the outcome of the observations. The positive results of the observations were, of course, decisive in convincing other physicists that he was right.

General relativity is the prime example of a physical theory built on a mathematical "leap in the dark." It might have remained undiscovered for a century if a man with Einstein's peculiar imagination had not lived. The same cannot be said of quantum mechanics, the other major achievement of twentieth-century physics. Quantum mechanics was created independently by Werner Heisenberg and Erwin Schrödinger, working from quite different points of view, and its completion was a cooperative enterprise of many hands. Nevertheless, in quantum mechanics too the decisive step was a speculative jump of mathematical imagination, seen most clearly in the work of Schrödinger.

Schrödinger's work rested on a formal mathematical similarity between the theory of light rays and the theory of particle orbits, a similarity discovered some 90 years earlier by the Irish mathematician William Rowan Hamilton. Schrödinger observed that the theory of light rays is a special limiting case of the theory of light waves that had been established after Hamilton's time by Maxwell and Hertz. So Schrödinger argued: Why should there not be a theory of particle waves having the same relation to particle orbits as light waves have to light rays? This purely mathematical argument led him to construct the theory of particle waves, which is now called quantum mechanics. The theory was promptly checked against the experimentally known facts concerning the behavior of atoms, and the agreement was even more impressive than in the case of the general theory of relativity. As often happens in physics, a theory that had been based on some general mathematical arguments combined with a few experimental facts turned out to predict innumerable further experimental results with unfailing and uncanny accuracy.

General relativity and quantum mechanics are success stories, showing mathematical intuition in a fruitful and liberating role. Unfortunately there is another side to the picture. Mathematical intuition is more often conservative than revolutionary, more often hampering than liberating. The worst of all the historic setbacks of physical science was the definitive adoption by Aristotle and Ptolemy of an earth-centered astronomy in which all heavenly bodies were supposed to move on spheres and circles. The Aristotelian astronomy benighted science almost completely for 1,800 years (250 B.C. to A.D. 1550). There were of course many reasons for this prolonged stagnation, but it must be admitted that the primary reason for the popularity of Aristotle's astronomy was a misguided mathematical intuition that held only spheres and circles to be aesthetically satisfactory.

Ptolemy explained the motions of the moon and planets by means of cycles and epicycles, that is to say, hierarchies of circles of various sizes, moving one on another. The devastating feature of Ptolemy's system was that, since it was tailored in detail to fit the observed motions of each planet, it was immune to observational disproof. By the time of Ptolemy (A.D. 150) the vital force of Greek mathematics had been extinguished, and there were no new mathematical ideas to contest the stranglehold of Euclid's spheres and circles on the scientific imagination. Disturbed neither by new celestial observations nor by new mathematics, the 1,000-year night set in.

When Kepler in 1604 finally demolished the epicyclic cosmology by his discovery that planetary orbits are ellipses, he was not helped by any mathematical preconceptions favoring elliptical motions. On the contrary, he had to fight tooth and nail against his own mathematical prejudices, which were still uncompromisingly medieval. Only after years of struggling with various systems of epicycles did he overcome his conservative tastes enough to consider a system of ellipses. Such mathematical conserv-

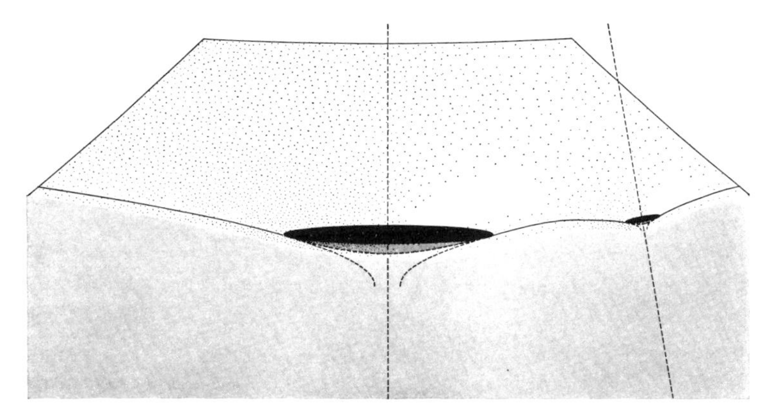

FIGURE 5. *Curvature of space was postulated by Einstein on the basis of very general arguments and aesthetic judgments. To build his theory Einstein used as his working material non-Euclidean geometry, a theory of curved spaces that had been invented during the 19th century. In this representation two massive bodies are shown in two dimensions on a two-dimensional surface. The local curvature of space around the bodies accounts for their gravitational properties. In actuality physical space-time is four-dimensional.*

atism is the rule rather than the exception among the great minds of physics. The man who breaks out into a new era of thought is usually himself still a prisoner of the old. Even Isaac Newton, who invented the calculus as a mathematical vehicle for his epoch-making discoveries in physics and astronomy, preferred to express himself in archaic geometrical terms. His *Principia Mathematica* is written throughout in the language of classical Greek geometry. His assistant Henry Pemberton, who edited the third edition of the *Principia*, reports that Newton always expressed great admiration for the geometers of ancient Greece and censured himself for not following them more closely than he did. Lord Keynes, the economist, who made a hobby of collecting and studying Newton's unpublished manuscripts, summed up his impressions of Newton in the following words:

"In the eighteenth century and since Newton came to be thought of as the first and greatest of the modern age of scientists, a rationalist, one who taught us to think on the lines of cold and untinctured reason. I do not see him in this light. I do not think any one who has pored over the contents of the box which he packed up when he finally left Cambridge in 1696 and which, though partly dispersed, have come down to us, can see him like that. Newton was not the first of the age of reason. He was the last of the magicians, the last of the Babylonians and Sumerians, the last great mind which looked out on the visible and intellectual world with the same eyes as those who began to build our intellectual inheritance rather less than 10,000 years ago. Isaac Newton, a posthumous child born on Christmas Day, 1642, was the last wonder-child to whom the Magi could do sincere and appropriate homage."

The character of Newton and his devotion to alchemy and to ancient apocalyptic writings are a fascinating subject, but it does not concern us here. We are concerned only with his mathematical style and tastes and with the effect of his mathematics on his science. Everything we know concerning his attitude toward mathematics is consistent with Keynes's conclusions. There is little doubt that Newton, like Kepler, made his discoveries by overcoming deeply conservative mathematical prejudices.

From these various historical examples we can only conclude that mathematical intuition is both good and bad, both indispensable to creative work in physics and also totally untrustworthy. The reasons for this two-edged quality lie in the nature of mathematics itself. As the physicist Ernst Mach remarked: "The power of mathematics rests on its evasion of all unnecessary thought and on its wonderful saving of mental operations." A physicist builds theories with mathematical materials, because the mathematics enables him to imagine more than he can clearly think. The physicist's art is to choose his materials and build with them an image of nature, knowing only vaguely and intuitively rather than rationally whether or not the materials are appropriate to his purpose. After the design of the theory is complete, rational criticism and experimental test will show if it is scientifically sound. In the process of theory-building, mathematical intuition is indispensable because the "evasion of unnecessary thought" gives freedom to the imagination; mathematical intuition is dangerous, because many situations in science demand for their understanding not the evasion of thought but thought.

I come now to discuss the present situation in physics. Without intending discourtesy to the experts in solid-state, nuclear spectroscopy and so forth, I use the word "physics" as an abbreviation for high-energy physics: the study of the fundamental particles. Physics (in this narrow sense) is now in an unusually happy situation. The latest generation of big accelerators has revealed during the past five years a whole new world of particles, with a quantity of detail and a richness of structure hardly anyone had expected. We must be profoundly thankful that the responsible physicists and politicians, not knowing that all these things were there, had the faith and courage ten years ago to go ahead with building the machines. As a result of their enterprise we now have a large amount of exact information about a world that is as new and strange as the world of atoms was in 1910. Just as in 1910, we have no comprehensive theory, and the theorists have complete freedom to make of the experimental data what they will.

In this situation the theoretical physicists choose their objectives and their methods according to criteria of mathematical taste. The primary question for a theorist is not yet "Will my theory work?" but rather "Is what I am doing a theory?" The material at hand for theoretical work consists of fragments of mathematics, cookbook rules of calculation and a few general principles surviving from earlier days. What combination of these items would deserve the name of a theory is a question of mathematical taste.

The three main methods of work in contemporary theory are called field theory, *S*-matrix theory, and group theory. They are not mutually exclusive; at least there is no contradiction between the things that adherents of the different methods do, although there is sometimes a contradiction between the things they say. Probably all three points of view will in the end make fruitful contributions to the understanding of nature.

The three methods differ not only in their choice of mathematical material but also in the uses to which the material is put. Field theory begins from a prejudice in favor of mathematical depth, a feeling that deep physical understanding and deep mathematics ought to go together. So the chosen mathematical material is the algebra of operators in Hilbert space, which is combined with various other difficult parts of mathematics in order to reach a structure that embodies some of the salient features of the real world. The emphasis is on a rigorous mathematical understanding of the theory, not on detailed comparison with experiment. Of the three methods, field theory is the remotest from experiment and the most mathematically strict, the most ambitious in its intellectual tone and the vaguest in its relevance to physics. I am myself addicted to it and am therefore particularly qualified to point out its limitations.

In *S*-matrix theory (the *S* stands for *Streu*, the German word for "scatter") the mathematical material is deliberately chosen to be as elementary as possible. It consists of the standard theory of analytic functions of complex variables, a theory whose essential features have not changed since it was created by the French mathematician Augustin Cauchy early in the nineteenth century. *S*-matrix theory compensates for the weakness of its mathematical base by making heavy use of experimental data. The *S*-matrix theoretician typically aims to compute or predict the result of one experiment by making use of the results of others. Sometimes predictions are made from "first principles" independent of other experiments, and the hope is ultimately to deduce everything from first principles. One of the most pleasant and refreshing features of *S*-matrix theory is that the rules of the game can be changed as a calculation proceeds. The method as it now exists is transitional; one is not applying a cut-and-dried theory but rather creating a theory as he goes along by a process of trial and error. At every stage of the work the comparison with experiment will ruthlessly eliminate the unfit idea and leave room for truth to grow.

The success of S-matrix theory in interpreting experiments, and in giving guidance to experimenters, has been impressive. My own preference for field theory is based on a personal taste that, judged by the evidence of history, cannot be considered reliable. I find S-matrix theory too simple, too lacking in mathematical depth, and I cannot believe that it is really all there is. If the S-matrix theory turned out to explain everything, then I would feel disappointed that the Creator had after all been rather unsophisticated. I realize, however, that He has a habit of being sophisticated in ways one does not expect.

I shall now discuss group theory, the third of the principal methods used in modern theoretical physics, in somewhat greater detail than the other two. The mathematical material here is a theory of considerable depth and power, mostly dating from the first quarter of the twentieth century. The two main concepts are "group" and "representation." A group is a set of operations possessing the property that any two of them performed in succession are together equivalent to another operation belonging to the set. For example, the three-dimensional rotation group O_3 is defined as the set of all rotations of an ordinary three-dimensional space about a fixed center. Obviously, if R_1 and R_2 are any two such rotations, the combination of R_1 with R_2 can be duplicated by a third rotation, R_3. A representation of a group is a set of numbers and a rule of transformation of these numbers such that each operation of the group produces a well-defined transformation of the numbers. The transformations in a representation are restricted to being linear; that is to say, if a particular transformation sends p to p' and q to q', then it also sends $p + q$ to $p' + q'$. An example of a representation of O_3 is the set of three coordinates (x, y, z) that determine the position in space of any point P (see Figure 6). When a rotation R is applied, the point P moves to a new position P' with coordinates x', y', z', and this determines the rule of transformation for x, y, z. This particular representation of O_3 is called the triplet representation, since there are three numbers involved in it.

The immense power of group theory in physics derives from two facts. First, the laws of quantum mechanics decree that, whenever a physical object has a symmetry, there is a well-defined group (G) of operations that preserve the symmetry, and the possible quantum states of the object are then in exact correspondence with the representations of G. Second, the enumeration and classification of all well-behaved groups and of their representations have been done by the mathematicians, once and for all, independently of the physical situation to which the groups may be applied. From these two facts results the possibility of making a purely abstract theory of the symmetries of fundamental particles, based on the abstract qualities of groups and representations and avoiding all arbitrary mechanical or dynamical models.

The crucial transition from concrete to abstract group theory is most easily explained by examples. An atom floating in a rarefied gas has no

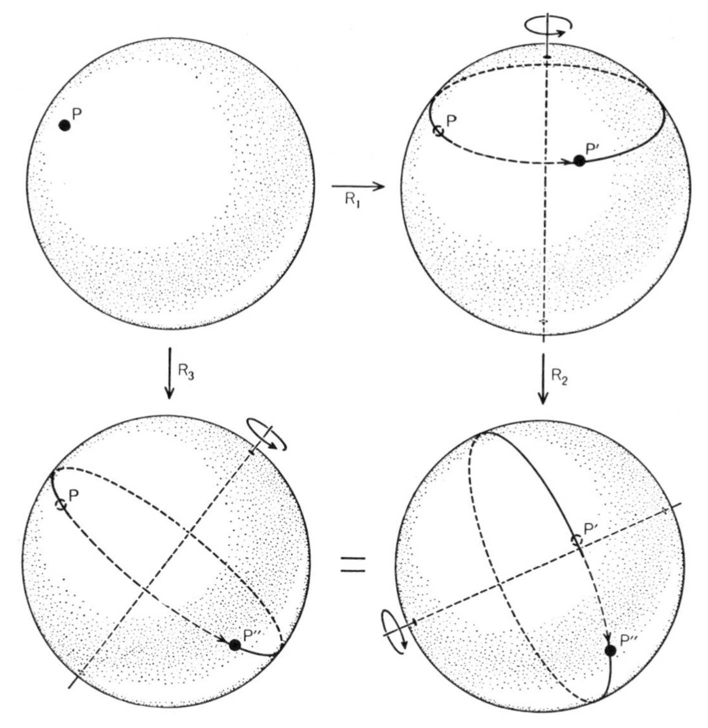

FIGURE 6. *Three-dimensional rotation group* O_3 *is defined as the set of all rotations of an ordinary three-dimensional space about a fixed center. If* R_1 *and* R_2 *are any two such rotations, the result of combining the two can be duplicated by a third rotation,* R_3.

preferred direction in space and therefore has the symmetry of the ordinary rotation group O_3. Among the representations of O_3 there is the triplet representation. Those states of the atom that have one unit of spin belong to this representation and are called triplet states; they always occur precisely in groups of three with the same energy. Now let a magnetic field be turned on so as to destroy the rotational symmetry; the three equal energies are slightly split apart and the three states can be seen in a spectroscope as a visible triplet of spectral lines. Such a classification of states of the atom according to their rotational symmetry is the standard example of concrete group theory at work.

Now we jump to a different example. There are three kinds of fundamental particle called pions, one positively charged, one negatively charged, and one neutral. They all have approximately the same mass and approximately the same nuclear interactions. Let us then imagine that they are a triplet representation of a group O_3', having exactly the same abstract structure as O_3 but having nothing to do with ordinary space rotations. We can then predict many of the properties of pions from abstract group theory alone without knowing anything about the intrinsic nature of the operations constituting O_3'. It turns out that all of these predicted properties of pions are correct. What is much more remarkable, these predictions were made on the basis of abstract group theory by Nicholas Kemmer in 1938, nine years before the first pion was discovered. The group O_3' (with some slight modification) is known in physics as the "isotopic-spin group."

Finally we come to the eightfold way, which gave us the key to the classification of the more recently discovered particles. The classification depends on a group U_3, which is larger and less familiar than O_3. To make U_3 understandable to nonmathematicians I shall introduce a mechanical model that bears the same relation to the abstract group U_3 as the rotations in three-dimensional space bear to the abstract group O_3. Needless to say, this mechanical model is not supposed to exist in the real world. It is intended only to illustrate the structure of U_3.

Consider a solar system in which the force of gravity varies directly with the first power of distance instead of with the inverse-square law. Suppose the planets to be small, so that their mutual perturbations are negligible. Each planet then moves independently in an elliptical orbit with the sun at the center. The peculiar feature of these orbits is that they all have the same period, the outer planets moving faster than the inner ones. We call the period of each orbit a "year," so that the positions of all planets repeat themselves at yearly intervals.

The motion of a planet can be specified precisely by two points in space denoted (P, Q), P being the position of the planet now and Q being the position it will occupy three months later. Another planet traveling three months ahead of the first in the same orbit will be specified by $(Q, -P)$, where $-P$ means the point diametrically opposite P. The total energy of either of these planets is given by $(OP^2 + OQ^2)$, which is the sum of the squares of the distances of the points P and Q from the sun at O. The group U_3 (as exhibited by this particular model) is defined as the set of all transformations of the planetary motions, subject to the following three restrictions: (1) the transformations are linear; (2) the transformations leave the total energy of each motion unchanged, and (3) if two or more planets are moving in a given orbit, a transformation that takes one to a new orbit takes all.

If only the first two conditions were imposed, we would have the group of all linear transformations of (P, Q), leaving the sum $(OP^2 + OQ^2)$ unchanged. This would be simply the rotation group O_6 in a space of six dimensions (three dimensions for P and three for Q). The group U_3 is thus a subgroup of O_6. The third restriction on U_3 can be stated in a more concise but equivalent form as follows: A transformation that takes the motion (P, Q) into (R, S) must also take $(Q, -P)$ into $(S, -R)$.

Two special kinds of transformation can easily be seen to belong to U_3. First, consider ordinary rotations operating on P and Q simultaneously. These obviously satisfy the three conditions. Hence the rotation group O_3

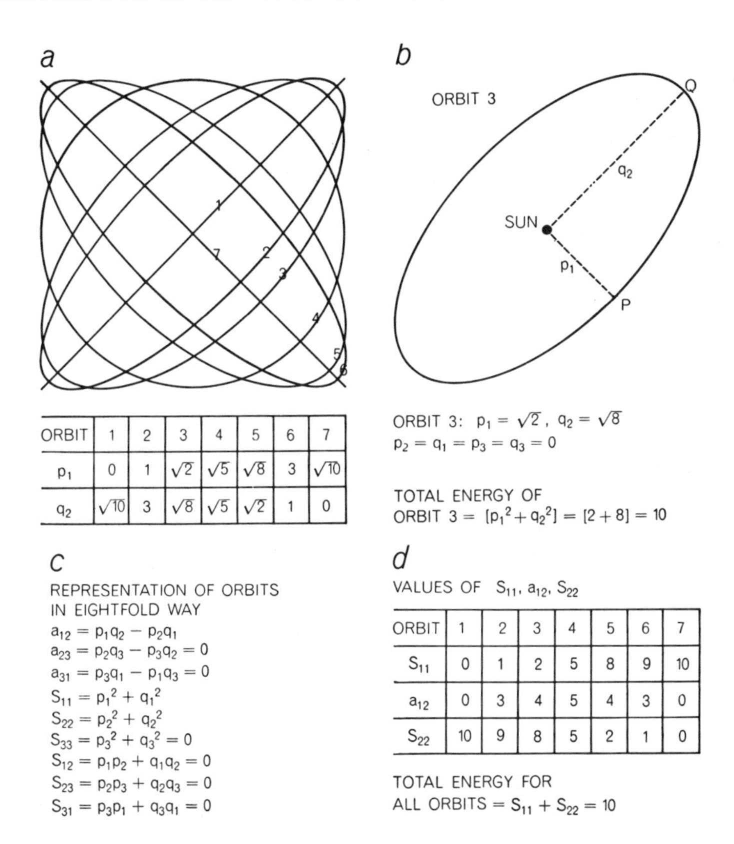

ORBIT	1	2	3	4	5	6	7
p_1	0	1	$\sqrt{2}$	$\sqrt{5}$	$\sqrt{8}$	3	$\sqrt{10}$
q_2	$\sqrt{10}$	3	$\sqrt{8}$	$\sqrt{5}$	$\sqrt{2}$	1	0

ORBIT	1	2	3	4	5	6	7
S_{11}	0	1	2	5	8	9	10
a_{12}	0	3	4	5	4	3	0
S_{22}	10	9	8	5	2	1	0

FIGURE 7. *Eightfold-way model bears the same relation to the abstract group* SU_3 *as rotations in three-dimensional space* (see FIGURE 6) *bear to the abstract group* O_3. *The model* (a) *shows seven planetary orbits that can be transformed into each other by operations belonging to the group* SU_3, *discussed in the text. That there are seven orbits is not significant; any number of others could be specified to satisfy the needs of this particular model. Orbit No. 3 is shown separately* (b) *to indicate how a planetary motion is defined by the points* P *and* Q, *with values* p_1 *and* q_2. *Normally six coordinates (three of* p *and three of* q) *are needed to define a point in space. But because of the special way the coordinate axes are chosen for this model,* p_2 *and* q_1 *are zero, and because the orbits all lie in a plane, the coordinates* p_3 *and* q_3 *are also zero. All the orbits have the same total energy* ($p_1^2 + q_2^2 = 10$) *but different angular momenta, expressed in terms of the value* a_{12}. *In particular the two straight-line orbits* (1 and 7) *have zero angular momentum, whereas the circular orbit* (4) *has the most angular momentum. According to the eightfold way the seven orbits can be represented by sets of nine numbers, listed in* c. *It is evident that six of these numbers vanish because* p_2, p_3, q_1, *and* q_3 *are all zero. Thus only three components remain:* S_{11}, a_{12}, S_{22}. *When the appropriate values for* p_1 *and* q_2 *are inserted, the three components take the values shown in* d. *The values are such that they transform into each other when the operations of* SU_3 *symmetry are applied. This is possible, in part, because the total energy for all orbits is the same:* $S_{11} + S_{22} = 10$.

is a subgroup of U_3. Second, consider the transformations of time displacement, in which each planet is transformed into the same planet at a fixed interval of time earlier or later. The time displacements also belong to U_3 and form another subgroup (T) of U_3.

For application to physics it is convenient to reduce U_3 to a smaller group SU_3 (and when I write SU_3, I mean the group the professionals call SU_3/Z_3). One obtains SU_3 from U_3 by simply forgetting time. For SU_3 all motions belonging to the same orbit are regarded as identical, irrespective of the time. Whereas U_3 transforms a planetary motion into another planetary motion at a particular time, SU_3 transforms an orbit into an orbit without reference to time. In mathematical language SU_3 is the group U_3 with the subgroup T of time displacements removed from it. The representations of SU_3 are precisely those representations of U_3 that are independent of time.

Let us now look for simple representations of SU_3. A planetary motion (P, Q) is defined by the six coordinates $(p_1, p_2, p_3, q_1, q_2, q_3)$ of P and Q. These coordinates by themselves define a representation of U_3 but, since they are time-dependent, not of SU_3. The simplest time-independent quantities (for reasons we need not go into) consist of p's and q's multiplied together in various combinations, as shown in Figure 7. There are nine and only nine such quantities. Three of them are components of the angular momentum and are designated a_{12}, a_{23}, and a_{31}; six are components of another kind that are related to the total energy of the system: S_{11}, S_{22}, S_{33}, S_{12}, S_{23}, S_{31}. (The subscripts indicate which of the p's and q's are involved in defining the quantity; thus $a_{12} = p_1q_2 - p_2q_1$ and $S_{12} = p_1p_2 + q_1q_2$.)

Because this representation of SU_3 involves nine quantities, it is said to be nine-dimensional. The sum $S_{11} + S_{22} + S_{33}$, however, is the total energy of the system and does not transform at all under any of the operations of U_3. When three quantities yield a constant sum, it is evident that only two of them are independent and that the third is always implied when any two are given. The three quantities S_{11}, S_{22}, and S_{33} might be reduced to two independent quantities in various ways, but for technical reasons they are usually combined as follows: One quantity is expressed as $S_{11} - S_{22}$ and the other as $S_{33} - \frac{1}{2}(S_{11} + S_{22})$. As a result there are only five independent components of S rather than six, and these together with the three components of a yield a total of eight quantities that do in fact transform into each other under U_3. These eight quantities are time-independent and provide an eight-dimensional representation of SU_3. The representation so obtained is the simplest that exists and is the famous eightfold way.

Finally, let us imagine that the SU_3 symmetry in nature is not perfect. Suppose, for example, that the "three direction" (that is, the direction in which the coordinates have the values p_3 and q_3) is somehow different from the other two directions. In terms of our imaginary solar system this

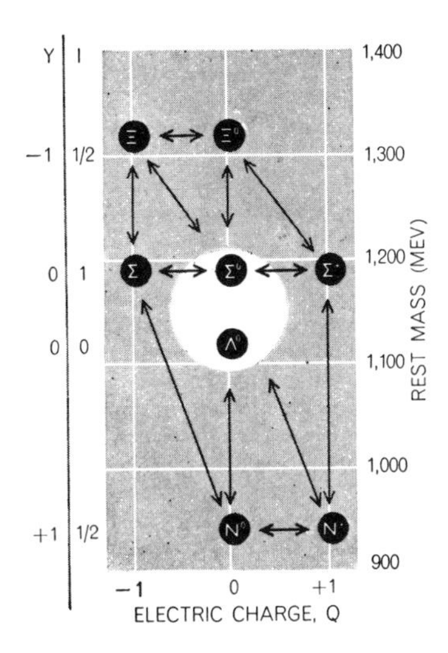

FIGURE 8. *Superfamily of eight was the first grouping suggested by the eightfold way. It contains the eight most familiar baryons: the neutron* (N^0) *and proton* (N^+) *— also known as the nucleon doublet — the lambda* (Λ) *singlet, the original sigma* (Σ) *triplet and the original xi* (Ξ) *doublet. The sigma triplet and xi doublet that appear in the 10-member superfamily containing the omega minus* (see Figure 1) *are heavier particles with the same values of* Y *and* I.

means that symmetry is preserved only when the orbits all lie in the same plane and is not preserved by rotations that carry the orbits off the plane. In this case the symmetry group U_3 will be replaced by its subgroup U_2, consisting only of those transformations of U_3 that leave the three direction unaltered. Under the operations of U_2 the eightfold way does not remain a unified representation. Its eight components split into subsets in the following manner:

$$S_{33} - \tfrac{1}{2}(S_{11} + S_{22}) \quad \text{(a singlet)},$$

$$\left.\begin{matrix} S_{23},\ S_{31} \\ a_{23},\ a_{31} \end{matrix}\right\} \quad \text{(two doublets)},$$

$$(S_{11} - S_{22}),\ S_{12},\ a_{12} \quad \text{(a triplet)}.$$

Each of these subsets forms a representation of U_2. In other words, the transformation defined by the singlet representation pertains to a unique subset of the total set: a subset of one member. Each of the two doublets represents a slightly larger subset: a subset of two members. Similarly, the triplet subset has three members.

Turning now to the actual physical world, compare this eight-member structure with the eight original baryons, the most familiar of the heavy "elementary" particles, which consist of the lambda (Λ) singlet, the proton–neutron (or nucleon) doublet, the xi (Ξ) doublet and the sigma (Σ) triplet. The agreement is exact.

In other representations of SU_3 there are 10, 27, or more members. Gell-Mann was the first to point out that the symmetry of a 10-member set could be satisfied by 9 of the known baryons if they were augmented by a missing singlet that he named, in advance, the omega-minus (Ω^-) baryon. The known members of this 10-member set were a delta (Δ)

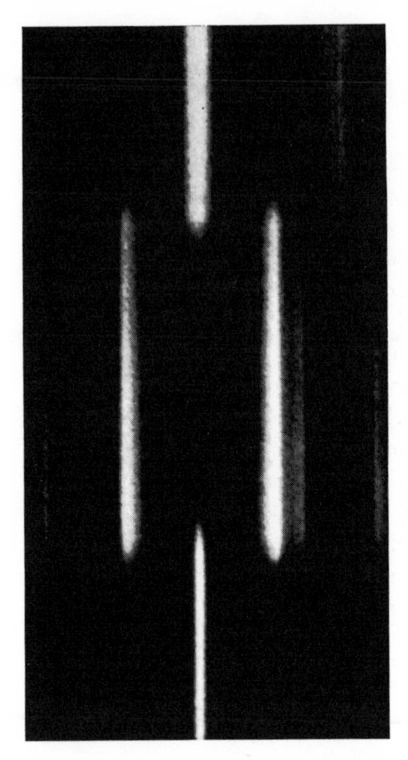

FIGURE 9. *Spectral line of niobium* (bottom) *is split into three components when a magnetic field is turned on so as to destroy the rotational symmetry of the atoms. Two components are observed perpendicular to the magnetic field* (middle) *and a third is observed parallel to the magnetic field* (top). *The triplet lines correspond to three states of the atom that have one unit of spin each; these states always occur precisely in groups of three with the same energy. Such a classification of states of the atom according to their rotational symmetry is an example of applied group theory. Spectrograms were made in the Spectroscopy Laboratory of the Massachusetts Institute of Technology.*

quartet, another sigma triplet and another xi doublet. The predicted singlet was discovered in February in bubble-chamber photographs made at the Brookhaven National Laboratory.

The evidence is now overwhelming that an abstract symmetry with the structure of SU_3 actually exists in nature and dominates the behavior of the strongly interacting particles. The symmetry is not perfect, being broken by some relatively weak perturbation that reduces the group SU_3 to its subgroup U_2. The U_2 symmetry that remains is essentially identical with the abstract isotopic-spin symmetry discussed earlier. Our entire picture of the strongly interacting particles has been transformed from chaos to a considerable degree of order by these compellingly simple group-theoretical ideas.

Group theory is in many respects the most satisfactory of the three theoretical methods I have discussed. Unlike S-matrix theory, it has an elegant and impeccably rigorous mathematical basis; unlike field theory, it has clear and solid experimental support. What then is lacking? The trouble with group theory is that it leaves so much unexplained that one would like to explain. It isolates in a beautiful way those aspects of nature that can be understood in terms of abstract symmetry alone. It does not offer much hope of explaining the messier facts of life, the numerical values of particle lifetimes and interaction strengths — the great bulk of quantitative experimental data that is now waiting for explanation. The process of abstraction seems to have been too drastic, so that many essential and concrete features of the real world have been left out of consideration. Altogether group theory succeeds just because its aims are modest. It does not try to explain everything, and it does not seem likely that it will grow into a complete or comprehensive theory of the physical world.

We are left with three methods of work in theoretical physics: field theory, *S*-matrix theory, and group theory. None of them really deserves the name of theory, if we mean by a theory something similar to the great theories of the past, for example general relativity or quantum mechanics. They are too vague, too partial, or too fragmentary. This is of course only my personal judgment. Even if they succeed in their declared aims, they do not satisfy my aesthetic sense of what a theory ought to be. I am tempted to apply to them the term "Bridges of snow built across crevasses of ignorance" to describe my feelings of dissatisfaction. This splendid phrase is often useful for characterizing theoretical ideas with which one happens to be unsympathetic. It is well to remember, however, that it was first so used by the bigoted biometrician Karl Pearson in a diatribe against Gregor Mendel's laws of inheritance.

BIBLIOGRAPHY

Arthur Koestler, *The Sleepwalkers*, Grosset and Dunlap, Inc., New York, 1963.

Geoffrey F. Chew, Murray Gell-Mann, and Arthur H. Rosenfeld, "Strongly Interacting Particles," *Scientific American*, *210*, No. 2, 74–93, February 1964.

Eugene P. Wigner, "The Unreasonable Effectiveness of Mathematics in the Natural Sciences," *Communications on Pure and Applied Mathematics*, *13*, No. 1, 1–14, February 1960.

BIOGRAPHICAL NOTE

Freeman Dyson is a professional physicist who has been educated and trained as a pure mathematician. As a student in England he was influenced by Hardy, Littlewood, and Besicovitch, and he almost became a number theorist. However, he later discovered that his particular mathematical skills were more appreciated by physicists than by pure mathematicians. So he switched to physics and has lived happily ever after, looking for problems in physics which could be fruitfully attacked by a retired mathematician. As a result, he has worked from time to time in many parts of physics, trying to demonstrate that mathematical elegance and physical truth are not always incompatible.

The mathematical theory of analytic functions has come to play an essential role in the theory of elementary particles in collision. Experiments produce probability curves describing the chance that one thing or another will happen when two or more particles collide, for example, a certain new particle will be produced and will emerge from the target, heading in a particular direction with a particular energy. The theoretical physicist seeks laws that correspond to these empirical curves. Subtle mathematical analysis of the theoretical curves probes into such fundamental physical questions as the role of causality in the subatomic world. "Do these concepts provide a framework within which an essentially complete and unique description of the structure and collisions of elementary particles is to be constructed?" the author asks.

Analytic Functions and Elementary Particles

A. S. Wightman

It is best to begin by describing some typical experiments in elementary particle physics and how they are analyzed theoretically. The customary experimental arrangement is illustrated in Figure 1. The accelerator

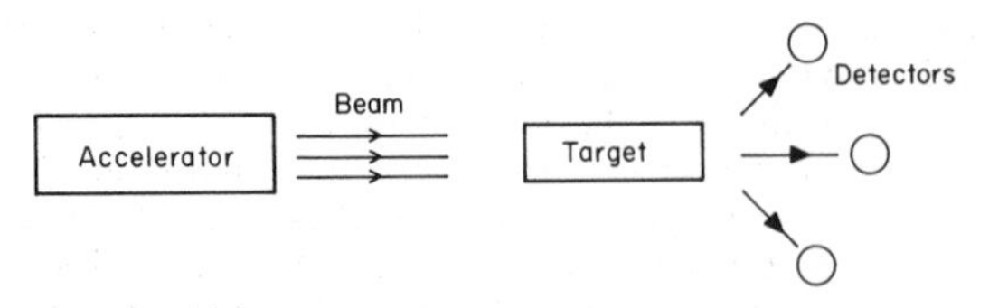

FIGURE 1

produces a beam of particles, typically protons or electrons, which then collide with a stationary target. The products of the collisions are either scattered particles or new particles. (One of the main advantages of accelerators which produce a high-energy beam is that the production of new particles is possible and likely.) In practice, the target and detectors may be combined in one device. For example, in a hydrogen bubble chamber the hydrogen atoms are targets, and both the charged particles of the beam and the charged particles produced give rise to visible tracks of bubbles in the liquid, which permit one to determine their directions of motion.

The particles in the beam are essentially independent of one another and, to a large extent, so are those of the target. Thus, the collisions can be regarded as a succession of independent events. They are customarily symbolized as shown in Figures 2a through 2c. Here the legs with arrows

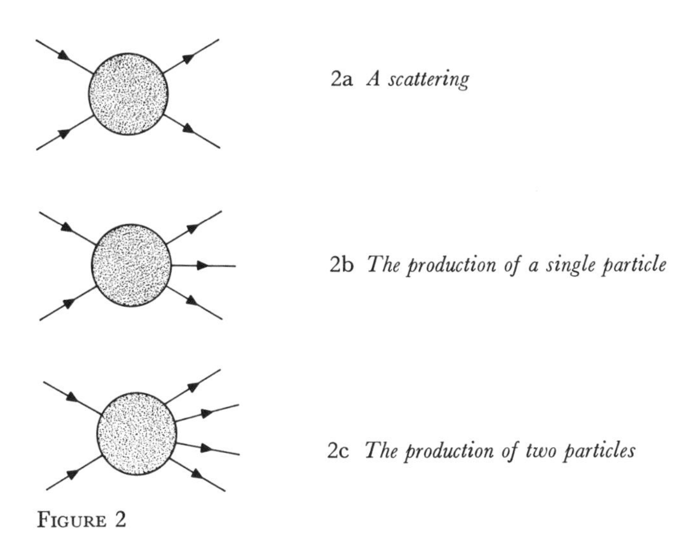

2a *A scattering*

2b *The production of a single particle*

2c *The production of two particles*

FIGURE 2

going into the box from the left denote colliding particles, while legs with arrows pointing out of the box to the right denote particles coming out of the collision. The black box in the middle is a suitable symbol for the collision itself. By convention, collisions proceed from left to right.

Sometimes the legs in the symbol for a collision are given a distinctive character to indicate different types of particles. For example, if ∼∼ denotes a photon and —— an electron, then Figure 2d is the scattering of light by an electron, so-called Compton scattering.

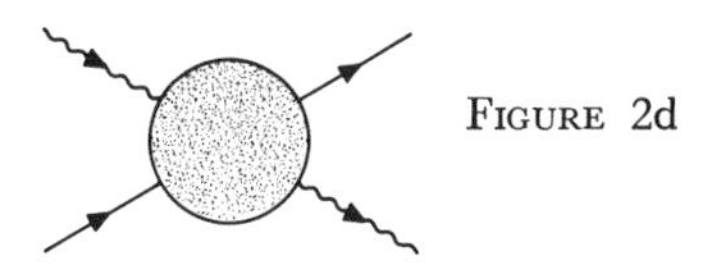

FIGURE 2d

This description of a collision is the appropriate one for an observer at rest in the laboratory. On the other hand, for an observer moving with the speed of the beam particles, it is the target particles that are in motion. One can think of intermediate cases, in which both target and beam particles are moving. This last situation is realized in practice in recent "colliding beam" experiments, in which two beams pass through each other, heading in nearly opposite directions. The symbolic notation for collisions just described fits well with all those possibilities. To get a quantitative description, one has merely to give the energies of the particles described by the legs on the diagram and the angles their directions of motion make with one another. Of course, these quantities will be measured only to within a certain precision, so that the description of a particle's energy will not, in fact, be that it has the value E but, rather, that it lies between E and $E + \Delta E$. Similar statements hold for measured angles.

It should be mentioned that, in addition to the collisions described so far, there are others that play an important role in theory, although in the laboratory they are not customarily directly observed because they turn out to be very infrequent. Examples are collisions symbolized by Figure 2e, referred to as three- or four-body collisions, respectively.

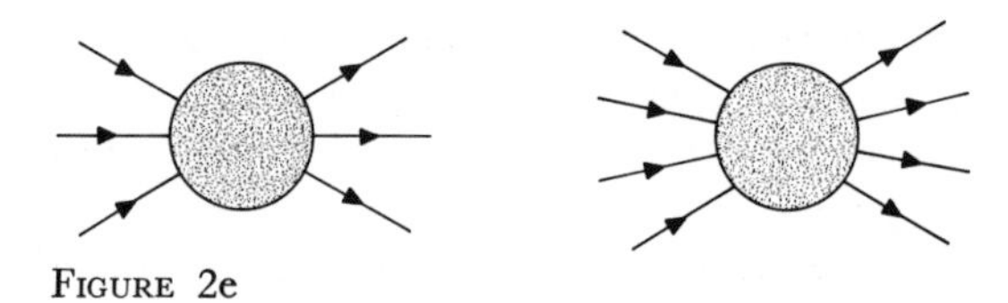

FIGURE 2e

This completes the description of an individual collision. An experiment consists of a statistical study of the frequencies of the different kinds of collisions. Thus, the typical outcome of an experiment on the scattering of two particles will be a family of so-called *histograms*, giving the frequency of collisions in which a particle is scattered through an angle between θ and $\theta + \Delta\theta$ for beam particles of energies between E and $E + \Delta E$ (Figure 3).

A theory is asked to produce a set of probability distributions to be compared with these histograms. The experimental curves are never

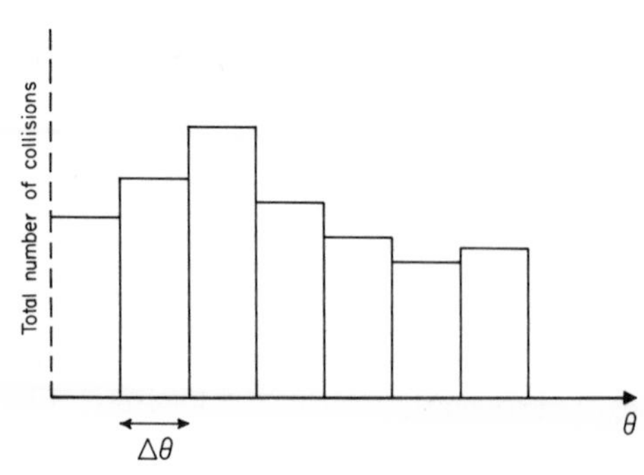

FIGURE 3

smooth. Of course, by refining the apparatus and running the accelerator longer, one can decrease $\Delta\theta$ and increase the number of events observed so that the experimental curve would gradually become a better and better approximation to a smooth curve. That is just what normally happens as experiments get better and better. It is this smooth curve that the theory describes.

In the theoretical descriptions of collisions, according to quantum mechanics, the probability distributions are predicted to be proportional to the square of the absolute value of a complex function $F = f_1 + if_2$ of the (real) variables that describe the experiment, say E and θ; F is called the "probability amplitude."

Piecewise Smooth and Piecewise Analytic Functions

After this long preliminary, I come to a mathematical question: What kind of function would one expect to have to use to describe these probability distributions? When an experiment is rough, this question is not very interesting. One has a scraggly histogram and is content to approximate it with some elementary function, say a linear or quadratic function. An example is Figure 4.

As experiments are refined, however, the previous question becomes more reasonable. To put it more specifically: Is it a reasonable *a priori*

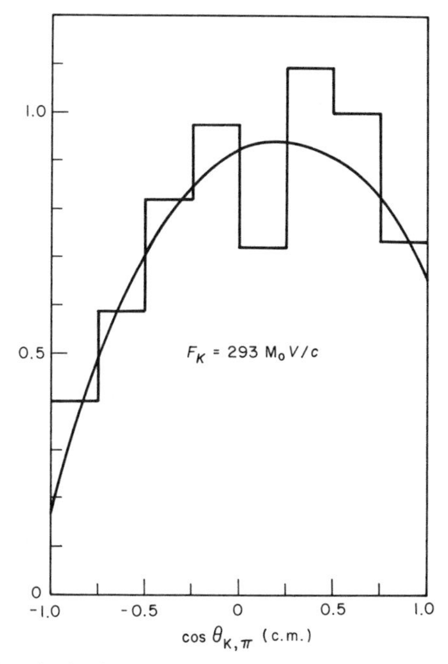

FIGURE 4. *The angular distribution for the absorption reactions* $K^- + p \rightarrow \Sigma^- + \pi^+$.

requirement that probability distributions predicted by a theory be smooth functions?

When the question is posed this way, it is easy to see that the answer is "No," because such a curve can have a "break" in it. For example, it can happen that a production process like that of Figure 2b has zero probability as a function of incident beam energy below some threshold energy. The probability curve may then have a "break" at the threshold as shown in Figure 5. The break in this case is associated with a discontinuity

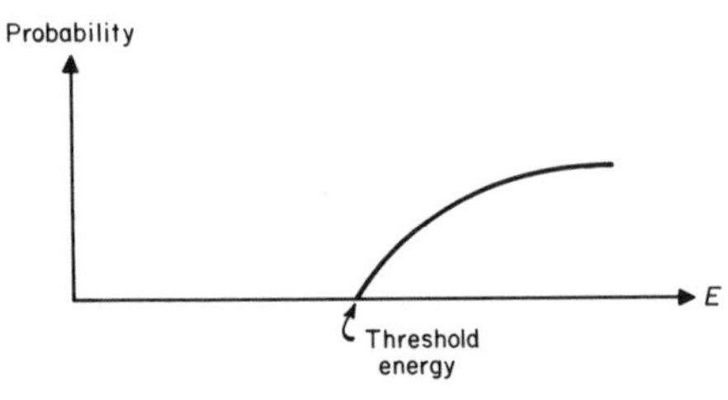

FIGURE 5

in the slope of the curve. In other cases, it might be associated with a discontinuity of some higher derivative or a discontinuity in the function itself. To accommodate this (and numerous other) examples, it is reasonable to change the adjective "smooth" in the question to "piecewise smooth," that is, for a function of one variable, smooth with the exception of some points where something special happens. The assumption now can be interpreted as saying that probability distributions ought to be smooth when there is not some physical reason to the contrary, and that will happen only rarely. If this is admitted, one is left with another question: How smooth is smooth? A straightforward answer is: A function is smooth at a point E if it has derivatives of all orders at E. It is smooth in an interval $E_1 < E < E_2$ if it is smooth at each point of the interval. (Such a definition has an evident extension to functions of several variables. For simplicity of exposition, I will often discuss the case of a single variable even though the case of several variables is the important one in practice.)

Now the customary assumption in elementary particle physics is not that probability distributions are piecewise smooth in this sense but that they are piecewise *analytic*, and thereon hangs our tale.

A function f, a real variable, is analytic at a point E_0 if it is smooth there and if the Taylor series for the function converges to the function in some neighborhood of E_0; that is,

$$f(E) = f(E_0) + f^{(1)}(E_0)(E - E_0) + f^{(2)}(E_0)\,\frac{(E - E_0)^2}{2!} + \cdots \quad (1)$$

for E in some interval around E_0 say $|E - E_0| < \delta$. Here $f^{(n)}(E_0)$ is the nth derivative of f at the point E_0. The requirement that the series be convergent turns out to be equivalent to a boundedness condition on the derivatives at E_0:

$$|f^{(n)}(E_0)| < C\,R^n\,n!, \tag{2}$$

for all positive integers n and some constants C and R independent of n. There are plenty of functions that are smooth without being analytic because, given any sequence of real numbers $a_0, a_1, a_2, \ldots,$ there exists a function smooth in a neighborhood of E_0 such that $f^{(n)}(E_0) = a_n$, and one can certainly choose the a_n so as to violate the boundedness Condition 2.

It is, therefore, natural to ask: On what grounds, experimental or other, does one restrict attention to piecewise analytic functions? There is an unsatisfactory answer which can be put aside immediately although it may be historically relevant: The fact that there are piecewise smooth functions that are not piecewise analytic is not widely known among physicists since such functions are rarely discussed in the traditional elementary mathematics course. A justification of the assumption of analyticity by a direct experimental test of the Inequalities 2 for the successive derivatives at a point has certainly never been carried out and looks entirely impractical. One can justify the assumption of analyticity either by its indirect consequences or by relating it to some other physical hypotheses. Both kinds of argument are important and will be discussed in turn.

Indirect Consequences of Analyticity

To bring out some of the indirect consequences of analyticity, I assemble some elementary facts about analytic functions. If the Taylor series (1) converges for all real numbers E satisfying $|E - E_0| < \delta$, it also does so for all complex numbers E satisfying the inequality. For a complex number of the form $E = E_1 + iE_2$, the inequality is $|E - E_0| = [(E_1 - E_0)^2 + E_2^2]^{\frac{1}{2}} < \delta$. If E_1 and E_2 are used as coordinates in the plane, this is a circle with center at E_0 on the real axis and radius δ, a so-called circle of convergence, as shown in Figure 6. The probability amplitude $f_1 + if_2$

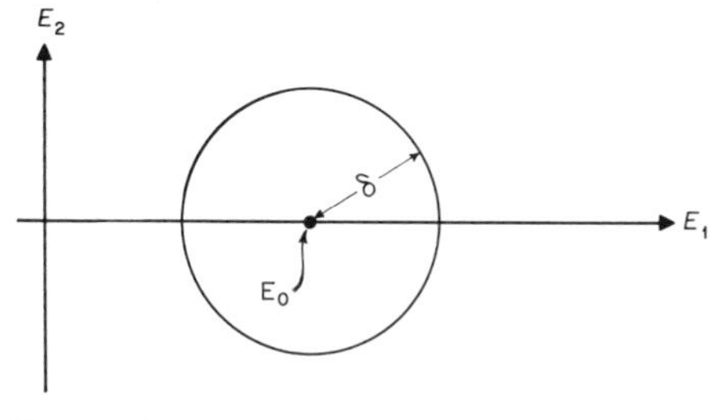

FIGURE 6

then is defined by the power series as a complex-valued function of the complex variable $E = E_1 + iE_2$. It is with such complex analytic functions that most elementary particle theory works these days.

Clearly, if all the derivatives of f_1 and f_2 as functions of the real variable E_1 vanish at a point E_0 of the real axis, the power series for f_1 and f_2 vanish in any circle of convergence. Furthermore, if $f_1 + if_2$ vanishes on any interval of the real axis, all derivatives of f_1 and f_2 vanish at each point inside the interval. These remarks constitute a derivation of the basic fact that an analytic function which vanishes on an interval of the real axis vanishes in any circle of convergence with center in that interval. This statement is also true if the function vanishes on any smooth curve inside a circle of convergence. It accounts for a second basic fact about analytic functions of a complex variable: If f is an analytic function with a circle of convergence C, and C_1 is another circle overlapping C, there is at most one way of extending the domain of definition of f to C_1 so that it is analytic there. For, if there were two extensions f_1 and f_2, their difference would be analytic and would vanish in the overlap between the two circles and, therefore, throughout C_1 (see Figure 7). The process of ex-

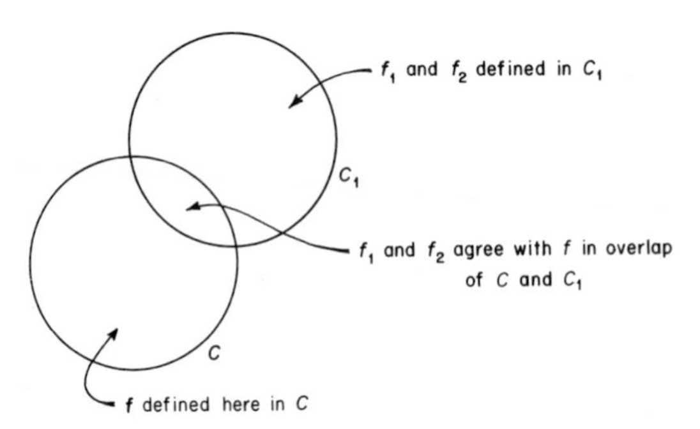

FIGURE 7

tending the definition of f in this manner is known as "analytic continuation." What prevents the extension of an analytic function everywhere by analytic continuation? The answer is *singularities* that prevent the convergence of the Taylor series. For example, the function $1/E$ can be continued analytically throughout the complex E plane with the exception of the point $E = 0$, where there is a singularity of the function.

The notion of analytic continuation is, in general, completely inapplicable to a function that is smooth but not analytic. Such a function need not be uniquely determined by its Taylor series, which may, in fact, diverge. It can vanish on a subinterval of an interval where it is smooth without vanishing everywhere.

In the application of these results of analytic function theory to collision theory, one needs rules that give a physical interpretation to probability amplitudes obtained by analytic continuation. These rules are either part of the basic assumptions of the physical theory in question or are to be deduced from other physical principles of the theory. For some situations the rules are almost self-evident. If one continues analytically along the real axis without encountering a "break" in the probability distribution curve for the collision in question, one expects the analytically continued amplitude to be identical with the physical probability amplitude. There are other situations where the rules are far from obvious and in fact their formulation represents a remarkable physical discovery. This is the situation with probability amplitudes related by what is customarily called *crossing symmetry*. In the simplest case, the path of analytic continuation leads from E_0 to $-E_0$ via the path shown in Figure 8. Now, the amplitude

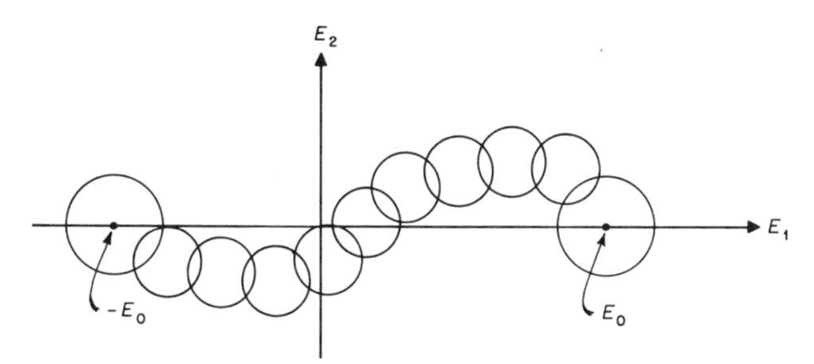

FIGURE 8

at $-E_0$ has no obvious physical meaning. The rule says it is to be related to the probability amplitude for a collision involving an antiparticle. To state precisely which collision that may be, I go back to the production reaction (Figure 2b). What is asserted is that analytic continuation in the variables associated with the leg encircled with the dotted line in Figure 9a leads to an amplitude whose symbol (Figure 9b) denotes a collision of two particles and an antiparticle to produce two particles. The backward arrow on the leg denoting the antiparticle is a reminder that the amplitude for the collision with an antiparticle energy of energy $+E$ is an analytic

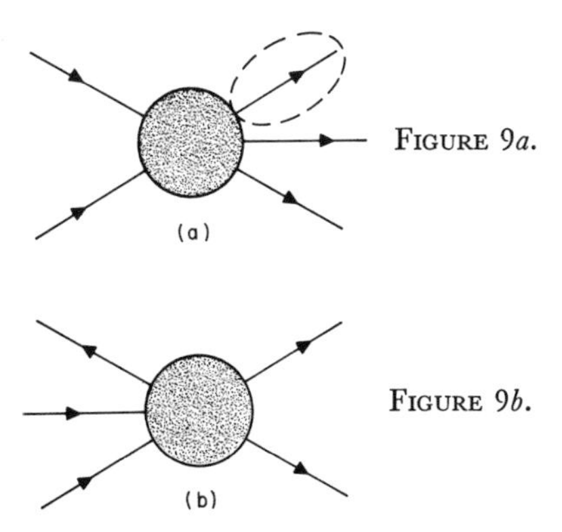

FIGURE 9*a*.

FIGURE 9*b*.

continuation to $-E$. (Also, the direction of motion has to be reversed, but we will ignore such fine points.)

To see how spectacular the predictions of crossing symmetry can be, one need only examine the symbol (Figure 2d) for Compton scattering, which describes the scattering of a photon by an electron. By a double application of the continuation associated with crossing symmetry, one deduces that the probability amplitude for the annihilation of an electron positron pair to produce two γ rays (Figure 10) is the analytic continuation

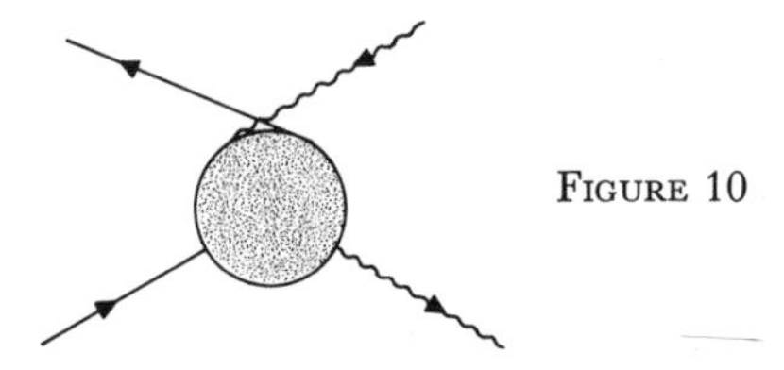

FIGURE 10

of the probability amplitude for Compton scattering. Although limited, all experimental evidence accumulated so far supports the view that crossing symmetry holds in Nature.

It seems to me that crossing symmetry alone gives sufficient grounds for the advisability of exploring the assumption that the probability amplitudes for collisions are piecewise analytic functions. For this reason, instead of expanding further on the successes of what is usually called dispersion theory, I turn to arguments for the assumption that relate it to other physical principles.

One of the most common tools for producing analytic functions in physics is the Laplace transform. It is defined as follows: If f is a function of a real variable t decent enough to be integrated (though not necessarily piecewise smooth) and vanishing for negative t, that is,

$$f(t) = 0 \qquad \text{for} \quad t < 0,$$

then

$$F(q) = \int_{-\infty}^{\infty} e^{-qt} f(t)\, dt = \int_{0}^{\infty} e^{-qt} f(t)\, dt$$

is its Laplace transform. Here F turns out to be analytic in q for positive q and can be continued analytically into the entire right half plane in the variable $q + ip$ (Figure 11). The essential point is that, for $q > 0$, the

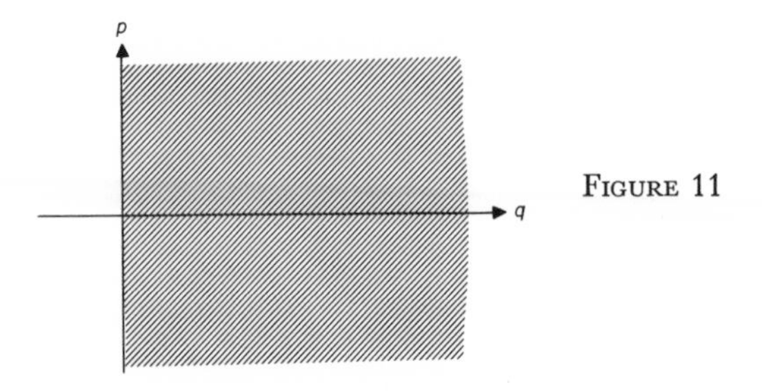

FIGURE 11

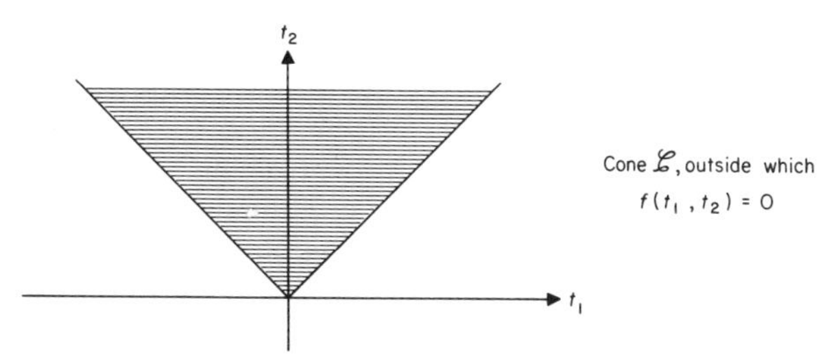

FIGURE 12

exponential e^{-qt} falls so rapidly to zero as $t \to \infty$ that the integral can be multiplied by any power t^n and still converge splendidly.

An analogous result holds in higher dimensions where t is replaced by $t_1, \cdots, t_n$. There the vanishing of f on the half line is replaced by the vanishing of f outside some cone. For example, in two dimensions, we have Figure 12, and the Laplace transform becomes

$$\begin{aligned} F(q_1, q_2) &= \int_{-\infty}^{\infty} \int_{-\infty}^{\infty} e^{-(q_1t_1+q_2t_2)} f(t_1, t_2)\, dt_1\, dt_2 \\ &= \int_{t_1} \int_{t_2 \text{ in cone } \mathcal{C}} e^{-(q_1t_2+q_2t_2)} f(t_1t_2)\, dt_1\, dt_2 \end{aligned}$$

This will again converge splendidly as long as $q_1t_1 + q_2t_2 > 0$ for all t_1, t_2 in the cone $\mathcal{C}$. The integral then defines an analytic function of two complex variables $z_1 = q_1 + ip_1$ and $z_2 = q_2 + ip_2$. The discussion is quite analogous for more variables.

To see how this construction arises in the physics of elementary particles, consider a function which measures some quantitative aspect of the correlation between two measurements at points P_1 and P_2 at times t_1 and t_2, respectively, with $t_2 > t_1$. (I am not concerned with which particular aspect; all that is essential for the moment is that, when there is no correlation, the function vanishes.) We admit the principle of relativity, which says that it is impossible to send signals faster than the speed of light. This means that, if P_2t_2 is out of the striped region shown in Figure 13, it cannot be affected by the measurement at P_1t_1; therefore, f vanishes outside the cone. This requirement is called *causality,* and it implies that

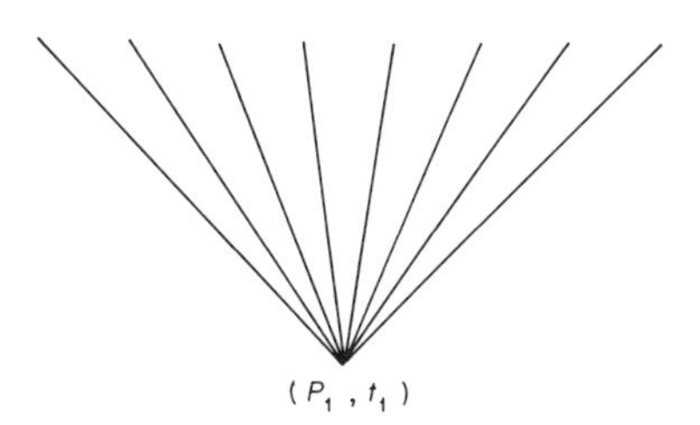

FIGURE 13

the Laplace transform of such a correlation function f defines an analytic function, provided that f is a decent function.

The significance of these analytic functions is that they are closely related to the probability amplitudes for collisions discussed before. In fact, it is one of the main concerns of modern quantum field theory to develop this relation systematically. Thus, within this framework of measurements in space and time, we have at least a rough explanation of the piecewise analyticity of collision amplitudes: It arises from the impossibility of sending signals faster than the speed of light.

This connection displays yet another aspect of the subtle and fundamental properties of physical theory that center on the uncertainty principle. The description of the possibilities of measurement in space-time regions associated with causality requirements is complementary to that in which the momenta of particles are measured. The Laplace transformation described earlier enables one to pass from the description of one set of possibilities to the description of another.

The mathematical tools of the theory of analytic functions of several complex variables used in working all this out are of rather recent vintage, so that there has been some fruitful interaction of mathematicians and physicists in this area, and there promises to be more.

Connection Between Collisions with Different Numbers of Legs

Crossing symmetry provides a connection between probability amplitudes for collisions with a given number of legs. It is natural to ask whether there are other relations that connect diagrams with different numbers of legs. One important family of such relations arises from the "principle of unitarity," which expresses the conservation of probability in collisions. How these come about can be understood as follows. In a collision, the incoming beams may be capable of producing a variety of reaction products. The theory describes the probability of these various possibilities as the square of the moduli of appropriate complex analytic functions. If we write down an equation expressing that the total probability of something happening in a collision is one, we have a relation which, in general, connects probability amplitudes for collisions with different numbers of legs. By considering all possible incoming beams, we arrive at an infinite set of relations among integrals of products of analytic functions. When combined with crossing symmetry, these relations are very restrictive. It is one of the most challenging problems of current theoretical physics to invent a theory of these relations, that is, to characterize the families of analytic functions that can describe the collisions of a given family of particles. So far there are only fragmentary results, but there are enough to show one characteristic feature of such solutions. We cannot tinker with one of the analytic functions involved without changing them all; the theory is all of one piece.

To recapitulate, the physical requirement of causality leads to the condition that analytic functions are involved in the mathematical formulation of describing the collisions of elementary particles. The analytic functions describing different collision processes are related by analytic continuations, resulting in crossing symmetry and other observationally verifiable results of great generality. The self-consistency of the physical interpretation of the mathematical formulation leads to unitary conditions. The complete mathematical structure under the combined requirement of unitary and crossing symmetry is so involved and subtle that, so far, only partial explorations have been made. Do these concepts provide a framework within which an essentially complete and unique description of the structure and the collisions of elementary particles is to be constructed? We do not yet know.

BIBLIOGRAPHY

C. N. Yang, *Elementary Particles*, Princeton University Press, Princeton, N.J., 1962.

K. W. Ford, *The World of Elementary Particles*, Blaisdell, Waltham, Mass., 1963.

Geoffrey F. Chew, Murray Gell-Mann, and Arthur H. Rosenfeld, "Strongly Interacting Particles," *Scientific American*, *210*, No. 2, February 1964, 74–93.

BIOGRAPHICAL NOTE

A. S. Wightman, a professor at Princeton University, is a member of both the mathematics and the physics departments. He explains his dual interest as follows: "As a student of elementary particle theory, I realized that its mathematical structure is of great subtlety and complexity. Since then, along with like-minded people, I have used everything but the kitchen sink in trying to understand that structure."

Sometimes regarded as an assortment of useful but rather dull tricks, numerical analysis is a vital branch of applied mathematics. Besides being the key to the effective use of electronic computers, it is an intellectually challenging field in its own right. It is now taught at different levels from high school to graduate school, and it draws on techniques and theories from many parts of mathematics. Still, it remains something of an art. The usual goal of numerical analysis is to get an answer that is good, though not necessarily perfect. Often it is difficult to recognize a good answer or, given a choice, to tell which of two answers is more satisfactory from a practical standpoint. Despite the sophistication of some of its methods, numerical analysis offers opportunities for ingenious people without a great deal of mathematical training. Plenty of work remains to be done to catch up with the brute processing power of today's computers.

Numerical Analysis

Philip J. Davis

Numerical analysis is the science and the art of obtaining numerical answers to certain mathematical problems. Some authorities claim that the "art" of numerical analysis merely sweeps under the rug all the inadequacies of the "science." It is simultaneously a branch of applied mathematics and a branch of the computer sciences.

This admittedly vague definition will be made clearer if we consider a typical instance of numerical analysis. A problem (let us say in physics) has led to a system of ordinary differential equations for the variables $u_1(t), u_2(t), \cdots, u_n(t)$. This system is to be solved, subject to the conditions that $u_i(t_0) = a_i$, $u_i(t_1) = b_i$, $i = 1, 2, \cdots, n$. This is the so-called two-point boundary-value problem. A casual inspection of the problem has revealed that there probably is no elementary closed-form expression that solves the problem, and it has therefore been decided to proceed numerically and to compute a table of values, $u_i(t_j)$, $j = 1, 2, \cdots p$, $i = 1, 2, \cdots, n$, which will be accepted as the solution. Numerical analysis tells us how to proceed.

The proper procedure may very well depend on the mechanical means for computation at our disposal. If we have pencil and paper, and perhaps a desk calculator, we should proceed along certain lines. If we have an electronic digital computer, these may be different lines. If the computer has certain features of memory and programming or if certain software is available, these may suggest economics of procedure.

Computation on a digital computer involves replacing the continuous variables $u_i(t)$ with discrete variables. But this can be done by a variety of methods. Will the differential equation be replaced by a difference equation? If so, this also can be done in many ways. What, then, is an appropriate way? If we adopt a finite difference policy, we shall be led to systems of ordinary equations that may be linear or nonlinear, depending on the original differential equation. How shall these equations be solved? Can we use direct methods? Or are we compelled to successive approximations via iterative methods? Yet again, and following a totally different approach, will the solution functions $u_i(t)$ be represented in some way by a finite combination of elementary functions, which has been determined so as to satisfy as well as possible the differential equation and the boundary conditions? We are confronted with many possible ways to proceed. Numerical analysis will have a word to say about each. Each separate mode of procedure is known as an "algorithm."

Having obtained some kind of answer to the problem by means of an algorithm, numerical analysis attempts to set bounds on how far this answer can differ from the true but unknown answer. Since we are working only approximately, errors can enter from a variety of sources. Barring blunders[1] (that is, machine malfunctions, erroneous programming, and other human errors), errors will arise from the fact that continuous variables have been made discrete, that infinite mathematical expressions or processes have been made finite or truncated, and that a computing machine does not do arithmetic with infinite exactitude but with say 8 figures. Numerical analysis attempts to make an error analysis for each algorithm. There are great difficulties with this type of problem, and the resulting bounds, when obtainable, may be deterministic or of a statistical nature. They may be *a priori* bounds or *a posteriori* bounds, that is, bounds that can be computed fairly simply prior to the main computation or bounds requiring that the whole computation be carried out beforehand. They may be mathematically demonstrable bounds or only approximate or asymptotic bounds. Finally, they may be bounds that the computer itself has computed.

What Is a Good Answer?

As part of error analysis, numerical analysis must even consider how to recognize a good answer when it meets one. What is the criterion of a good answer? There may be several. As a very simple example, suppose

[1] Blunders should not be dismissed so lightly. They occur with sufficient frequency so that the practicing numerical analyst must learn how to recognize them and how to deal with them.

that we are solving a single equation $f(x) = 0$ and that x^* is the mathematically exact answer. An answer $\bar{x}$ has been produced by a computation. Is $\bar{x}$ a good answer if $|x^* - \bar{x}|$ is small? Or is it a good answer if $f(\bar{x})$ is approximately zero? Considering that we probably cannot compute f exactly but can compute only an approximate $\bar{f}$, should we perhaps say that $\bar{x}$ is a good answer if $\bar{f}(\bar{x})$ is approximately zero? Various criteria may lead to widely differing answers.

Having obtained a reasonable answer to a problem, we may be interested in improving the answer. Can we obtain answers of increasingly good quality via a fixed type of algorithm? Numerical analysis has information on this, and the resulting theory, known as convergence theory, is one of the major aspects of the subject.

Numerical analysis therefore comprises the strategy of computation as well as the evaluation of what has been accomplished. The beau ideal of the subject (not often achieved, unfortunately) consists of

1. The formation of algorithms.
2. Error analysis, including truncation and roundoff error.
3. The study of convergence including the rate of convergence.
4. Comparative algorithms, which judges the relative utility of different algorithms in different situations.

Numerical analysis ranges continuously from theoretical mathematics on the one side to programming on the other. It is not possible to delineate its exact boundary.

The ultimate measure of success of the field must be the extent to which numerical analysis succeeds in obtaining solutions of demonstrable accuracy to problems and the extent to which these problems are, in fact, of great interest in other branches of mathematics or are known to embody adequate models of important phenomena of the real world.

A Very Brief History of Numerical Analysis

The origins of numerical analysis are ancient. The Babylonians developed arithmetic processes to a high degree of sophistication. Twelve hundred years before Pythagoras taught the square of the hypotenuse (that is, about 1700 B.C.), the Babylonians had very accurately approximated $\sqrt{2}$. The Archimedean (250 B.C.) inequality

$$3\tfrac{10}{71} < \pi < 3\tfrac{10}{70}$$

and its underlying theory, if slightly primitive, is completely modern in its statement and aims. The Table of Chords (trigonometric functions) of Klaudios Ptolmaios (A.D. 150) is an early masterpiece of numerical analysis and computation.

As mathematics developed and deepened after the twelfth century, the relevant numerical analysis also developed. There are many areas where the desire for numerical answers was the prime mover in work that subsequently turned out to be far more important to pure theory. Think of logarithms or of infinite series. There are consequently many areas of classical analysis that are inextricably tied up with the forward march of numerical analysis. It would be both fruitful and fascinating if someone would undertake a study of just where numerical analysis has, in fact, been a major motivating force.

As problems became more difficult, it became clear that brute computation was laborious, time-consuming, and very often carried out inaccurately. This led to the development of a variety of mechanical instruments of computation. Also, once the computational steps had been set down in the abstract, any particular computation could be carried out by men who could do brute computation accurately but who needed no over-all mathematical talent. The ultimate step is, of course, completely automatized computation via electronic computers.

As problems deepened, it became obvious that algorithms for their numerical solution were not easy to come by; in many instances the difficulty was very great. This led, on the one hand, to the build-up of theories to handle these difficulties and, on the other hand (several centuries later), to a mode of mathematical thought and argumentation which concentrated on the platonic existence and properties of solutions rather than on constructive, algorithmic methods. In the past several decades, under the stimulating necessity of getting numerical solutions, the process has once again been partially turned around, and constructive as opposed to existential methods have come somewhat more into the limelight.

Computation Before Electronic Computers

It is instructive to examine computing practice in the period 1935–1950. Mechanical adding machines and electric desk calculators capable of performing the simple arithmetic operations were available. In universities, they were generally to be found in the physics, the statistics, or perhaps the astronomy department. The mathematics department rarely had such a machine; it was more likely to have a set of Napier's rods and bones, exhibited in a glass case alongside a broken plaster hyperboloid. Slide rules of both the ten- and twenty-inch variety were available and commonly used, though seldom by mathematicians. In large laboratories, specialized mechanical devices such as planimeters were employed. Specialized laboratories may also have had electrical or photoelectrical equipment rigged up for specialized jobs such as Fourier analysis; they performed these jobs with indifferent success. Analog machines were in their infancy. Nomograms were commonly used in engineering offices.

But during this period, the solid production computation was performed by a person (frequently a woman) at a desk, with pencil and paper, a desk computer, a small library of fundamental mathematical tables, and a fairly small but time-honored collection of algorithms.

The computations were necessarily modest in scope. Most of them were the computation of combinations of elementary or slightly more complicated functions. Finding zeros of functions, the solution of differential equations, and so on, was much more laborious and was avoided if possible.

The outer limits of computation were rapidly reached. As examples we might cite the following. In 1942, systems of linear equations were rarely solved if the order exceeded 10 or 12. In the same period, it would have been a major effort to generate and find the zeros of the Legendre polynomials of order greater than 16.

Despite what in retrospect appears to be the limited accomplishments of the period, one fact must be borne in mind: The algorithms available then appear to have been commensurate with actual computing power. Today, when raw computing power has been multiplied by, say, a factor of 10^6, an estimated 80 percent[2] of all computations are done by using algorithms that were well developed prior to 1930.

Computation in the Period 1950–1967

The electronic computer multiplied our power of brute calculation by an enormous factor and combined arithmetic operations with a logical feature that adds tremendous flexibility to the programming process. A computation that previously might have extended over a period of several months (and hence was undertaken rarely, and even then with trepidation) is now performed in several minutes. We can now routinely invert a (100×100)-matrix, solve large systems of simultaneous differential equations, solve boundary-value problems of partial differential equations, and the like. Hardly an area of computation has not responded to the enormous multiplication of the calculation power. This great speed has meant that in numerous instances the computer may be linked up to keep pace with a physical process going on concurrently.

Side by side with this great multiplication of computer power, there has occurred a multiplication in the demand for computers. There is hardly a college worth its salt which does not now sport a computer center and a program of grants to support it. Some places where, in the old days, one would have been lucky to find a desk machine or a table of Bessel functions are now full of faculty and students from all departments, talking about hardware and software the way people used to talk about automobiles in the 1920's. In the near future, remote consoles and time-sharing arrangements will make the central computer omnipresent and

[2] Sheer guesswork. I have in mind computation carried out in a large computer center with a diverse business.

will have an impact on almost all branches of scientific education, from elementary computing to ways of doing homework.

The impact of all this on numerical analysis itself has been much more modest. Though enormously stimulated, both technologically and financially, numerical analysis has been moving forward slowly, in direct proportion, let us say, to the number of people who are in the field rather than to the number of computers spread over the country or to the ratio of the desk to electronic multiplication times. There are many computation centers with a full panoply of equipment which produce little or nothing new in numerical analysis (but, of course, produce useful results in other areas). There may be a parallel between this situation and what the small observatory has been in relation to astronomy. In the period 1850–1900, many colleges, at no little cost, fitted themselves out with small observatories. Not much astronomy ever came out of these 'scopes, but on pleasant evenings in the fall the public may still be admitted to take a peek.

The great speed of electronic computers has in some instances changed the relative value of well-established approaches. In some calculations, we may be able to do things most expeditiously by doing them in the unsophisticated way. For example, classical mathematics devotes many clever chapters to the problem of maximizing functions. In a real computation of maximization, it may turn out that it is easier simply to compute the function at a large number of stations and to pick out the observed maximum. Surprisingly, the machine has brought about here and there a return of mathematical primitivism as a viable mode of computational life.

The machine seems to have relegated nomograms and certain graphical methods to the memory book of mathematics. Although these subjects are frequently pretty and ingenious, they appear to play no great role in current numerical analysis.

Devices such as certain relaxation methods for differential equations, which depend on the eye and the experience of the person doing the computing, have lost favor to well-formulated algorithms.

In past generations, the difference calculus was of great importance in numerical analysis, both in the theory and in the practice. There has been a marked decrease in the use of differences in favor of the functional values themselves.

Numerical experimentation has been greatly stimulated.

In speaking of the rise and fall of certain types of mathematics, we must insist that no mathematical method ever be ruled out of consideration from computing practice. Changing machine characteristics and more versatile programming languages have restored to good practice numerous mathematical devices previously considered too cumbersome or costly.

One must also be alert to the possibility that pictorial input and output for the computer are going to be an essential part of the scene in the near future. This may mean that methods for handling partial differential equations (and other problems) that depend on the judgment of the human eye may make a strong comeback in the near future.

The well-prepared student who expects to do a Ph.D. thesis in numerical analysis should, therefore, be thoroughly conversant with linear algebra, complex variable, and probability theory and should know the elements of functional analysis. The research worker in the field will find these disciplines vital.

Thus, numerical analysis draws increasingly on the resources of pure mathematics. The reverse process is also going on; numerical analysis has in recent years suggested many interesting and deep problems in other branches of mathematics, for example, finite difference methods applied to partial differential equations are conveniently phrased in the language of matrices. This has led to the creation of new matrix theory as well as a revitalization of certain neglected portions. Approximation theory with great potential applications to computation has also been greatly stimulated. In some areas of approximation theory, computer practice is well ahead of adequate theoretical discussions. Numerical analysis has returned to pure mathematics the beautiful theory of Spline Approximation. It has deepened the theory of convex bodies. The theory of continued fractions, either long dormant or in the hands of a few specialists, has undergone development, and ignorance of the subject is not so profound as it once was. The theory of "computability," spurred by the existence of real (not just platonic) machines, has been returning dividends to mathematical logic.

Educational Opportunities

Prior to 1945, numerical analysis was a neglected area as far as formal courses are concerned. Few textbooks were available on the subject (the books by Whitaker and Robinson and by Scarborough come to mind), and formal courses were available at very few universities. The governing philosophy in those days seemed to be that numerical analysis consisted largely of a few tricks that one picked up as he needed them and that the intellectual level of the subject was low compared with other areas of mathematics.

World War II saw a vast amount of computing, followed by the development of many new techniques and by the writing of books. The electronic computer, which was born just as the war ended, lent tremendous impetus to numerical analysis.

The situation in 1967 is vastly different. The production of books in the field, both elementary and advanced, is nothing short of enormous. Most colleges and universities have a computer facility and, as an adjunct to this, have established undergraduate and graduate majors in computer science. Most colleges now offer at least one formal course in numerical analysis, and many universities offer four or more semesters work in numerical analysis at the graduate level. It is now possible in numerous universities to obtain the Ph.D. degree in numerical analysis. This degree may be granted by a department of mathematics, applied mathematics, or computer sciences.

The requirement of a course in computer sciences as part of an undergraduate major in science is becoming commonplace. As a result, practically all science majors will come in contact with numerical analysis directly or indirectly.

In high schools and in prep schools, numerical analysis is also increasing in popularity. Numerous high schools have either a computer on their premises or access to a nearby industrial or academic facility. Numerical analysis is being taught as a one-semester twelfth-grade option in some first-rate high schools, and textbooks on programming and numerical analysis appropriate to twelfth grade are appearing in increasing number.

Side by side with the movement to establish numerical analysis as a full-fledged course, there is also a growing tendency to introduce numerical analysis and computation in a meaningful way in traditional mathematics courses. Thus, for example, in calculus and advanced calculus one may employ the computer to investigate the concepts of functions, limits, infinite series, definite integrals, and so on. This tendency is bound to gain strength over the next ten years.

The whole area of numerical analysis in scientific education is currently in great flux, owing to the changing nature of physical facilities and software and also to the tastes of the instructor. Numerical analysis merges into programming on one side, into hardware on another side, into numerical calculus on still another side, and into algorithmic processes on a fourth side. Stressing one or more of these leads to different courses.

The Present State of the Field

The most casual observer of the field will see bustle, activity, and enthusiasm. More mathematical researchers than ever before spend part or all of their time in numerical analysis. More students than ever before are learning the elements of the field. More books on the topic, both elementary and advanced, textbook and research, have appeared than ever before (one would have to list approximately one hundred fifty relevant books in the last twenty years). The number of technical journals devoted principally to numerical analysis has increased from one in 1950 to about ten in 1966. In each monthly issue of *Mathematical Reviews* the number of reviews in the field averages about fifty.

The number of unsolved problems is large; the methods available for solution run the whole range from pure mathematics through programming, computer technology, and numerical experimentation. It is a field in which one can make a significant contribution with a modest amount of theoretical knowledge and in which one can bring to bear many of the resources of two thousand five hundred years of mathematics. Recent surveys have shown that perhaps up to one quarter of the college graduates who major in mathematics take jobs in the computer field and, hence, presumably will spend a portion of their professional life with numerical analysis.

This is a very cheerful picture. But there are a number of disquieting signs that should be mentioned. The term "numerical analysis" is, surprisingly, newly coined. It hardly appears before 1950. Prior to that, the words "approximate computation," "computational mathematics," "numerical mathematics," and the like, were used. The new term "numerical analysis" seemed to proclaim a new role and a new dignity for the subject. Yet ten years of this new dignity had hardly passed when this writer began hearing the question, "Is numerical analysis dead?" Of course, the game of asking the question, "Is X dead?" is a popular parlor game today. It helps increase the sale of picture magazines. But it is important for us to understand what underlies the special case X = numerical analysis.

Our subject appears to be torn by a number of opposing tendencies. On the one hand, there is an increasing tendency for the theoretical aspects of the subject to be pursued for their own sake, in isolation from the day-to-day needs of the man with serious computation problems. On the other hand, promoted by those who are fond of the computer as a piece of equipment, there has been a vast increase in what one friendly critic has termed "computerology." By focusing attention on programming, superprogramming, man-machine and machine-machine interplay, computerology has created its own problems, technical and theoretical, and devised its own languages and modes of communication. Though standing at a distance from mathematics itself, computerology requires some of the talents of mathematics and is invested with an abundance of glamour, so that a part of this talent has been drained off from the bluestocking field of numerical analysis. Degeneration sets in when computerology is pursued for its own sake. This is going on to an increasing extent. Will America's love affair with the automobile be replaced by one with the computer? This would be computerology's finest hour.

A related tendency comes from the feeling that it should be possible to employ "automatic programs" or methods of "artificial intelligence" so that any given problem in numerical analysis can be solved as perfunctorily as one now adds two digits. In other words, it should be possible to "automate out" numerical analysis. Whether or not this is a theoretically attainable goal, I leave for mathematical metaphysicians to work out. I simply record that at present the practicing programmer who expects to live entirely on such an elementary facility as a library of SHARE routines is in for a bit of a shock.

As long as applied mathematics itself is a live thing, providing new problems, new ideas, and new methods, numerical analysis will share this vitality.

BIBLIOGRAPHY

Several useful references are included in the bibliography following the next essay, entitled "Solving a Quadratic Equation on a Computer" by George E. Forsythe. In addition, Professor Davis himself has written three books with bearing on numerical analysis: *The Lore of Large Numbers*, Random House, New York, 1961; *Interpolation and Approximation*, Blaisdell, Waltham, Mass., 1963; *The Mathematics of Matrices*, Blaisdell, Waltham, Mass., 1965.

BIOGRAPHICAL NOTE

Philip J. Davis is Professor of Applied Mathematics at Brown University. After receiving his Ph.D. from Harvard University in 1950, he did postdoctoral research at Harvard's School of Engineering in the mathematical theory of compressible flows. He was then for a dozen years a member of the staff of the National Bureau of Standards, where he became Chief of the Numerical Analysis Section. He is especially interested in approximation theory and in the numerical application of the more abstract methods of functional analysis.

As every high school student knows, the solutions of a quadratic equation $ax^2 + bx + c = 0$ are always given by a single simple formula in terms of a, b, and c. From the standpoint of algebra, this completely settles the matter. From the standpoint of finding useful numerical approximations to these solutions, however, even the mightiest computers sometimes encounter pitfalls, and numerical analysts concerned with practical aspects of computing must use great ingenuity to prevent insignificantly small errors from snowballing and producing ridiculous answers. In view of the widespread use of computers, some practical aspects of computation should be part of everyone's basic education in mathematics, the author suggests.

Solving a Quadratic Equation on a Computer

George E. Forsythe

The automatic digital computer is one of man's most powerful intellectual tools. It forms an extension of the human mind that can be compared only with the augmentation of human muscle by the most powerful engines in the world.

Computers have a wide variety of applications, ranging from the control of artificial satellites to the automatic justification and hyphenation of English prose, and even to the storage and searching of medical libraries. However, computers were originally invented to carry out arithmetic computations rapidly and accurately, and this remains one of the major uses of computers today.

As early as World War I, L. F. Richardson indicated how the weather might be forecast with the aid of a vast computation then far beyond human capability, provided that enough upper-air weather observations were available as input data. By the 1940's the upper-air observations were beginning to appear in quantity. Hence, when John Von Neumann and others asked the government for funds to support computer development, they promised that computers would make it possible to carry out the arithmetic part of a modern version of Richardson's program. It was expected that the weather would soon be forecast routinely by computer, and so it was. It was even hinted that computers might make it possible to predict the future course of hurricanes after a variety of human interventions and thus lead to the eventual *control* of the weather. This, disappointingly, has not come to pass.

Many intellectual steps are involved in a project like weather forecasting by computer. In the first place, a reasonable model of the weather must be reduced to systems of equations, both algebraic and differential. The actual solution of such systems of equations is completely beyond the powers of any computer because, among other things, this involves knowing the wind at each of an infinite number of points of space.

Consequently, the second stage of numerical weather prediction is to replace the infinite number of points of space by a finite number of points arranged in a number of square-mesh layers spaced one above another, made of squares perhaps one hundred miles on a side. Instead of trying to describe and predict the wind at each point of space, one describes and predicts it at the corners of the meshes. The equations that describe the exact flow of air and moisture are replaced by much simpler equations, relating these quantities at neighboring points of the mesh. A great deal of mathematical analysis and experimental computation is needed before we can discover simple equations for the mesh which reasonably well simulate the actual equations for continuous space.

At the end of the second stage just described, we have a set of equations to solve. Each equation deals with unknown quantities that are real numbers. Recall that real numbers may be thought of as infinite decimal expansions like

$$-3.3333333333\ 3333333333\ 3333333333\ldots\ldots\ldots$$

(3's continued without end) or

$$3.1415926535\ 8979323846\ 2643383279\ldots\ldots\ldots$$

(digits continued without end, but without a predictable pattern). Since a computer is necessarily a finite collection of parts, it cannot hold all the decimal digits of even a single general real number. Hence, the third stage of the use of computers for weather prediction involves the use of a finite number system to simulate the real number system of mathematics.

Now that automatic digital computers have become so widespread in this nation, practically all of our scientific computation is done on them. The main point of this article is to bring out the difference between *theoretical* computation (that of arithmetic and algebra) and *practical* computation (that actually carried out on a computer).

We shall now describe the computer number system and some difficulties involved in using it. To illustrate the difficulties, we shall consider a mathematical problem that is almost infinitely simpler than the equations of meteorology — the quadratic equation.

Floating-Point Numbers

We shall first describe a simplified computer number system, the system of so-called floating-point numbers, and then show some of its behavior with a simple mathematical computation.

The usual number system of a computer reduces the infinite number of decimal places of real numbers to a fixed finite number. We first consider decimal numbers with a sign (+ or −) and one nonzero digit to the left of the decimal point, and exactly seven zero or nonzero numbers to the right of the decimal point. Examples of such numbers are −7.3456780, +1.0000000, +3.3333333, and −9.9808989. We say that such numbers have 8 *significant digits*. (The choice of 8 significant base-10 digits is purely for illustration; other systems also are used.) One can represent approximately 200,000,000 different numbers in this way, but they all lie between −10 and −1 or between +1 and +10. To enable computers to hold much bigger and smaller numbers, we add a sign and two more decimal digits to serve as an exponent of 10. In one commonly used system, the exponent is allowed to range from −50 to +49. Thus, the number $-87\frac{2}{3}$ is represented by

$$-8.7666667 \times 10^{+01}.$$

In this system, which is much like so-called *scientific notation*, the representable numbers all have 8 significant decimal digits. They range from

$$-9.9999999 \times 10^{49} \quad \text{to} \quad -1.0000000 \times 10^{-50}$$

and from

$$+1.0000000 \times 10^{-50} \quad \text{to} \quad 9.9999999 \times 10^{49}.$$

The number zero (0) is represented by $+0.0000000 \times 10^{-50}$. Approximately 20,000,000,000 distinct real numbers are thus representable in the computer, and these take the place of the infinite system of mathematical real numbers.

This computer number system is called the *floating-point number system*. The "point" is the decimal point. The exponent permits the decimal point effectively to "float" as much as 50 places away (to the left or the right) from its home position, corresponding to exponent 0.

Special computer facilities make it possible also to use so-called *double precision numbers* — numbers which have not 8 but 16 significant digits, with the exponent kept between −49 and +50. The machine works more slowly when these double precision numbers are used, but this penalty varies greatly among different computers.

In this paper we shall write floating-point numbers in various ways, but there will be understood to be exactly 8 significant digits. For example, we may write the number eleven as 11 or 11.0 or $+1.1000000 \times 10^{+01}$ or 1.1×10^1.

Computer Arithmetic

Besides holding floating-point numbers, every scientific computer must be able to perform on them the elementary arithmetic operations of addition and subtraction, multiplication and division. Let us consider addition first. Sometimes the exact sum of two floating-point numbers is itself a floating-point number; for example,

$$(+\ 2.1415922 \times 10^{+00}) + (+\ 9.7182818 \times 10^{+00}) \\ = +\ 1.1859874 \times 10^{+01}.$$

In this case, the computed sum is the same as the exact sum, and the computation is said to be without rounding error. More frequently, the exact sum is not a floating-point number. For example, the exact sum of $+6.6666667 \times 10^{+01}$ and $+6.6666667 \times 10^{+01}$ is 133.333334, a number with 9 significant digits. Hence, the exact sum cannot be held in the computer but must be rounded to the nearest floating-point number, in this case to $+\ 1.3333333 \times 10^{+02}$. This is a typical example where computer addition is only approximately the same as mathematical addition.

An even worse defect of computer addition appears when the numbers are numerically very large, so that the sum exceeds the capacity of the floating-point system. For example, the true sum of $+\ 9.9900000 \times 10^{+49}$ and $+9.9990000 \times 10^{+49}$ is 1.9989×10^{50}, a number greater than the largest possible floating-point number in our system. The computer should signal in some manner that an "overflow" has occurred and give the problem-solving program some option about what action to take. But it is impossible to store an answer that represents the exact sum to even one significant digit.

Computer multiplication suffers from the same two defects of computer addition — the necessity for rounding answers and the possibility of exponent overflow. While ordinary rounding errors are no more serious than with addition, overflow can be far worse, for the following reason. The exact sum of two floating-point numbers cannot exceed 2×10^{50}, but the exact product of two floating-point numbers can be as large as 9.9999998×10^{99}, and the product of two numbers each as small as 10^{25} can lead to overflow. Moreover, there is a possibility of "underflow" in multiplication. For example, the true product of 1.01×10^{-30} and 1.01×10^{-35} is 1.0201×10^{-65}, a number smaller by a factor of 10^{-15} than the smallest nonzero floating-point number. The most we can expect from the computer is that it will replace the product by zero and give the program a signal that underflow has occurred.

The problems encountered in computer subtraction are analogous to those in addition, and in this sense division is analogous to multiplication.

We assume that our computer operations of addition, subtraction, multiplication, and division, in the absence of overflow or underflow, will yield as an answer the floating-point number closest to the exact real answer. (In case of a tie, we permit either choice.) In fact many actual computer systems achieve this accuracy or come close to it.

Are Floating-Point Numbers Satisfactory?

Anyone who uses a digital computer for scientific computation is faced with a number system which is only approximately that of mathematics and arithmetic operations which are only approximately those of true addition, subtraction, multiplication, and division. The approximations appear to be very good, generally correct to within one unit in the eighth decimal digit. Only the most sophisticated of all scientific and engineering computations deal with numbers accurate to anything close to eight decimal places. We might therefore presume that rounding errors would provide no trouble in most practical computations. Moreover, the range of magnitudes from 10^{-50} to 10^{+49} safely brackets the range of all important physical and engineering constants, so that we might presume that we would have no trouble with overflow or underflow.

Is the floating-point arithmetic system so good that we can use it without fear to simulate the real number system of mathematics? Computer designers certainly hoped so and chose the numbers of significant digits and the exponent range with this expectation. However, the answer by now is clear: We may *not* proceed without fear. There are real difficulties. On the other hand, it is often possible to proceed with intelligence and caution and to get around the difficulties. But it has required an astonishing amount of mathematical and computer analysis to get around the difficulties, particularly in large problems. And, so far, we know well only how to handle relatively simple mathematical problems.

We shall illustrate some of the difficulties and their solution in the context of an elementary but important problem, the well-known quadratic equation of elementary algebra.

The Quadratic Equation

The reader will recall considerable time spent in the ninth grade, or thereabouts, finding the two roots of equations like

$$6x^2 + 5x - 4 = 0. \tag{1}$$

One first acquires some experience in factoring the quadratic. School examples *do* factor with a frequency bewildering to anyone who has done mathematics outside of school. For example, take

$$6x^2 + 5x - 4 = (2x - 1)(3x + 4). \tag{2}$$

If the left-hand side of Equation 2 is to be 0, then either $2x - 1$ or $3x + 4$ must be 0. The two possibilities tell us that the roots of Equation 1 are $\frac{1}{2}$ and $-\frac{4}{3}$.

Factoring in whole numbers is not always possible and turns out to be unnecessary, for one soon learns a formula that gives the two roots of any quadratic equation without any factoring. The main result is the following, the so-called quadratic formula.

If a, b, and c are any real numbers and if $a \neq 0$, the quadratic equation

$$ax^2 + bx + c = 0$$

is satisfied by exactly two values of x, namely,

$$x_1 = \frac{-b + \sqrt{b^2 - 4ac}}{2a} \tag{3}$$

and

$$x_2 = \frac{-b - \sqrt{b^2 - 4ac}}{2a}. \tag{4}$$

As an example of the use of the quadratic formula, the roots of Equation 1 are

$$x_1 = \frac{-5 + \sqrt{5^2 - 4(6)(-4)}}{12} = \frac{-5 + \sqrt{121}}{12}$$

$$= \frac{-5 + 11}{12} = \frac{6}{12} = \frac{1}{2}, \; x_2 = \frac{-5 - \sqrt{121}}{12} = -\frac{16}{12} = -\frac{4}{3}.$$

The roots agree, of course, with those found by factoring.

One great advantage of the quadratic formula is that it suggests a straightforward series of steps proceeding from the real numbers a, b, c to the solutions x_1, x_2.

Any such systematic process for computing some desired answer is called an "algorithm." In an algorithm no guesses are allowed — one proceeds directly from the data to the answer. The importance of algorithms is that computers have been expressly designed to be able to carry out algorithms and nothing but algorithms. That is to say, the logical steps performed by a computer are exactly those of an algorithm.

Next we give a detailed algorithm for evaluation of the quadratic formula (Equation 3). (It could be simplified.)

Algorithm for computing one root x_1 of the quadratic equation $ax^2 + bx + c = 0$.

(i) Compute $z = -b$.
(ii) Compute $y = b^2$.
(iii) Compute $w = 4a$.
(iv) Compute $v = wc$.
(v) Compute $u = y - v$.
(vi) Compute $t = \sqrt{u}$.
(vii) Compute $s = z + t$.
(viii) Compute $r = 2a$.
(ix) Compute $x_1 = s/r$.

Notes: 1. For simplicity, we assume here that $u = b^2 - 4ac$ is not negative, to avoid having to deal with imaginary numbers like $\sqrt{-1}$.

2. An algorithm for computing x_2 requires the replacement of steps (vii) and (ix) by

(vii)′ Compute $s' = z - t$.
(ix)′ Compute $x_2 = s'/r$.

In mathematics the above algorithm is implicitly understood to use real numbers and to carry out with them exact arithmetic operations including addition, subtraction, multiplication, division, and even extraction of the square root of $u = b^2 - 4ac$. As we showed earlier, a computer cannot carry out these exact arithmetic operations and, indeed, cannot even hold all real numbers a, b, c. Thus, although a real digital computer can carry out the exact logical steps of the algorithm, it must replace all numbers by floating-point numbers and all arithmetic operations by approximate operations.

The question, then, is this: Do the limitations of actual computer floating-point systems make any appreciable difference in solving quadratic equations?

The answer is: Sometimes yes and sometimes no. We shall give examples to illustrate both cases.

Examples of the Quadratic Formula on a Computer

Example 1 $$6x^2 + 5x - 4 = 0$$

For this example of Equation 1, the algorithm offers no difficulty to a computer with the precision we have given except, possibly, for the square root required in step (vi). Let us make the reasonable assumption that we have a method (indeed, another algorithm) for computing square roots with an error not exceeding 0.8 of a unit in the least significant decimal place. In this case, we will find $t = \sqrt{u} = \sqrt{b^2 - 4ac}$ to be 11.000000. Then we find that

$$x_1 = (-5 + 11.000000)/12 = 0.50000000,$$

a perfect result. The computation of x_2 leads to no loss of accuracy until the final division:

$$x_2 = -16.000000/12.000000 = -1.3333333,$$

as rounded on the computer. Since this is the correctly rounded value of the true x_2, we conclude that the computer algorithm has done as good a job as it could possibly do.

Example 2 $$x^2 - 10^5 x + 1 = 0.$$

Before examining the computer solution, we note that the true solutions, rounded to eleven significant decimals, are

$$x_1 = 99999.999990$$

and

$$x_2 = 0.000010000000001.$$

Moreover, it can be shown that x_1 and x_2 are well determined by the data in that small changes of the coefficients $(1, -10^5, 1)$ would cause only slight changes in x_1 and x_2.

Now let us apply the algorithm and see what are the computed values of x_1 and x_2. We have

$$\begin{aligned} a &= 1.0000000 \times 10^{+00}, \\ b &= -1.0000000 \times 10^{+05}, \\ c &= 1.0000000 \times 10^{+00}. \end{aligned}$$

First, to get x_1:

$$\begin{aligned} z &= -b = 1.0000000 \times 10^{+05}, \\ y &= b^2 = 1.0000000 \times 10^{+10}, \\ w &= 4a = 4.0000000 \times 10^{+00}, \\ v &= wc = 4.0000000 \times 10^{+00}, \\ u &= y - v = 1.0000000 \times 10^{+10}, \\ t &= \sqrt{u} = 1.0000000 \times 10^{+05}, \\ s &= z + t = 2.0000000 \times 10^{+05}, \\ r &= 2a = 2.0000000 \times 10^{+00}, \\ x_1 &= s/r = 1.0000000 \times 10^{+05}. \end{aligned}$$

The step that calls for comment is the computation of $u = y - v$, where the value of v is completely lost in rounding the value of u to eight decimals. The final answer x_1 is correct to eight decimals.

We now compute x_2:

$$
\begin{aligned}
z &= -b = 1.0000000 \times 10^{+05}, \\
y &= b^2 = 1.0000000 \times 10^{+10}, \\
w &= 4a = 4.0000000 \times 10^{+00}, \\
v &= wc = 4.0000000 \times 10^{+00}, \\
u &= y - v = 1.0000000 \times 10^{+10}, \\
t &= \sqrt{u} = 1.0000000 \times 10^{+05}, \\
s' &= z - t = 0, \\
r &= 2a = 2.0000000 \times 10^{+00}, \\
x_2 &= s'/r = 0.
\end{aligned}
$$

This time the computation of s' results in complete cancellation; so s' and, hence, x_2 are both 0. Thus, our algorithm has yielded a value of x_2 that differs by approximately 10^{-5} from the correct answer, and this might be considered a rather small deviation. On the other hand, our computed value of x_2 has a relative error of 100 percent — not a single significant digit is correct. Can this be considered a reasonable computer solution of the quadratic?

A study of ways in which quadratics are applied leads to the conclusion that the measure of accuracy should be that of *relative error*. As long as a root of a quadratic is well determined by the data and is in the range of floating-point numbers, a good algorithm should give it correctly to several or most of its leading digits, however large or small the root may be. Thus, we must conclude that the computer application of our quadratic algorithm for x_2 gave us practically no useful information about the root x_2. It follows that the algorithm is not adequate for solving a quadratic equation on a computer, because an adequate algorithm ought to work in *every* case.

Example 3 $\quad 6 \times 10^{30}x^2 + 5 \times 10^{30}x - 4 \times 10^{30} = 0.$

This is simply Example 1, with all its coefficients a, b, c multiplied by the factor 10^{30}. Thus, the roots are unchanged.

However, the algorithm breaks down at the second step, because $y = b^2$ is truly 2.5×10^{61}, a number outside the range of floating-point numbers. Therefore, the algorithm is again inadequate, though for a very trivial reason. A simple scaling of the coefficients (e.g., division of the equation by 6×10^{30}) would prevent the overflow.

Example 4 $\quad 10^{-30}x^2 - 10^{30}x + 10^{30} = 0.$

Here the true roots are extremely close to 10^{60} and 1. One of the roots is outside the range of floating-point numbers, and we could hardly expect to get it from a computer algorithm. The problem is: Can a reasonable computer algorithm get the root near 1?

Note that a simple scaling to make the first coefficient equal to 1 will cause the second and third coefficients to overflow. Hence, a scaling that helps cope with Example 3 will break down with Example 4. Certainly our algorithm will not work.

Does the reader feel that equations with such a large root will not occur in practical computations? Let him be assured that they do. The final, physically important result of a computation is almost certain to lie safely inside the range of floating-point numbers. Intermediate results, however, often have magnitudes smaller than 10^{-50} or larger than 10^{49}.

Recently several computing experts agreed that one of the most serious difficulties with many current computer systems is that they automatically replace an underflowed answer by zero, without any warning message. In such a system, $10^{-30} \times 10^{-30} \times 10^{31} \times 10^{30}$ would be computed as 0 (because underflow would occur with the first multiplication step), whereas $10^{-30} \times 10^{31} \times 10^{-30} \times 10^{30}$ would be correctly computed as 10.

Example 5 $\quad x^2 - 4.0000000x + 3.9999999 = 0.$

The correctly rounded roots are, to 10 significant digits,

$$x_1 = 2.000316228 \quad \text{and} \quad x_2 = 1.999683772.$$

If we apply the algorithm, we find that

$$\begin{aligned}
z &= -b = 4.0000000,\\
y &= b^2 = 16.000000,\\
w &= 4a = 4.0000000,\\
v &= wc = 16.000000,\\
u &= y - v = 0,\\
t &= \sqrt{u} = 0,\\
s &= z + t = z - t = 4.0000000,\\
r &= 2a = 2.0000000,\\
x_1 &= x_2 = s/r = 2.0000000.
\end{aligned}$$

The computed roots are both in error by approximately 0.0003162. That is, out of 7 computed digits to the right of the decimal point, only 3 are correct. Also, the computer mistakenly finds a double root instead of two different roots.

The accuracy seems quite low. However, the roots of Example 5 actually change very rapidly when the coefficients are changed. In fact, the two computed roots, $x_1 = x_2 = 2.0000000$, are the exact roots of the nearby equation $0.999999992x^2 - 3.999999968x + 3.999999968 = 0$. Thus, though x_1 and x_2 are wrong roots of Example 5 by some 3162 units in their last decimal place, they are true roots of an equation with a, b, c differing from those of Example 5 by no more than 0.8 of a unit in their last decimal place.

Example 5 illustrates two different ways of measuring relative errors in any computation. In the so-called "forward" approach to relative error, the computed roots x_1 and x_2 differ by a certain number of units (here 3162) in the last place from the true roots of the given equation. In the so-called "backward" approach to relative error, the computed roots x_1 and x_2 are the exact roots of an equation with coefficients which differ by no more than so many units (here 0.8) in the last place from those of the given equation. The forward measure of error is perhaps more natural and certainly traditional. The backward approach to error is more recent, but in many contexts turns out to be considerably easier to analyze and just as useful in practice. Backwards error analysis is one of the major research ideas to be developed in computational mathematics since the middle fifties. Cornelius Lanczos devised the backward approach in another context in the 1940's. Wallace Givens exploited it in 1954 for computing roots of certain equations. But James Wilkinson has done the most in the years since 1958 to exploit it as a basis for analyzing errors in floating-point computation on digital computers.

The reason that backwards error analysis is so useful is this: Some basic properties on which algebra is based (the associative and distributive laws) fail to hold for floating-point arithmetic. Hence, a forward error analysis, which is based directly on the floating-point operations, is extremely difficult to carry out. On the other hand, backwards error analysis interprets the result of each computer product, for example, as the true product of two real numbers that differ very slightly from the factors of the computer product. Thus, in backwards error analysis one deals with true mathematical multiplication and addition, which are associative and distributive. This permits analysis to be much more easily carried out and often leads to closer bounds for the error.

This is not the place to develop these ideas further, but we hope to have given the reader an inkling of why backwards analysis, when applicable, is often so much more satisfactory.

Criteria for a Good Quadratic Equation Solver

The previous examples illustrate the behavior and misbehavior of the usual quadratic algorithm. Examples 2, 3, and 4 make it clear that for computer application, the algorithm is not satisfactory for all cases and, hence, is unacceptable. What do we really expect from a quadratic equation-solving algorithm for use on a computer?

Should we be content with the computer solution of Example 5, with its error of 3162 units in x_1 and x_2, since the computed roots do satisfy an equation that is so close to the given one?

If we didn't know how to do better, we might be quite content with the results of Example 5 but certainly not with Example 2. Quadratic equations arise in many contexts of mathematics and computing. They are so basic that we should like to be able to compute their roots with almost no error for almost any equation whose coefficients are floating-point numbers. Such performance can be achieved, and it is essential to have such algorithms in the computer library. Then, when a quadratic equation occurs in the midst of a complex and imperfectly understood computation, we can be sure that the quadratic equation solver can be relied upon to do its part automatically and well, so that we can concentrate attention on the rest of the computation.

We want a quadratic equation solver that will accept any floating-point numbers a, b, c, and compute close decimal approximations to any of the roots x_1, x_2 that lie safely within the range of floating-point numbers. Any computed root should have an error in the last decimal place not exceeding, say, 10 units. If either x_1 or x_2 underflows, there should be a message about what happened.

Such an algorithm has been devised by William Kahan of the University of Toronto. The most difficult matter to take care of is the possibility of overflow or underflow. It will not be possible to describe the complete algorithm, but we can give some of the more accessible ideas.

First, we discuss the steps taken to overcome the great inaccuracy in root x_2, as computed in Example 2. In step (vii)′ of Example 2, we subtracted two equal numbers, z and t, to get $s' = 0$. The true value of t was not quite equal to z, because in step (v) the true u was not quite equal to y. But t and z, like u and y, could not be distinguished with only 8 decimal digits at our disposal.

An easy cure for the difficulty is to use another method of computing x_2 which does not involve the subtraction of nearly equal numbers.

If a, b, c are any real numbers, and if $a \neq 0$ and $c \neq 0$, then the quadratic equation

$$ax^2 + bx + c = 0$$

is satisfied by exactly two values of x, namely,

$$x_1 = \frac{2c}{-b - \sqrt{b^2 - 4ac}} \tag{5}$$

and

$$x_2 = \frac{2c}{-b + \sqrt{b^2 - 4ac}} \tag{6}$$

Formulas 5 and 6 can be proved, for example, by first applying Formulas 3 and 4 to the following equivalent form of the given quadratic equation:

$$c\left(\frac{1}{x}\right)^2 + b\left(\frac{1}{x}\right) + a = 0.$$

Note that if b is negative, there is cancellation in Formula 4 for x_2 but not in Formula 6, and there is cancellation in Formula 5 for x_1 but not in Formula 3. The reverse statements hold if b is positive. So, for any quadratic equation in which neither a nor c is zero, one selects Formulas 3 and 6 when b is positive or zero and Formulas 4 and 5 when b is negative. For Example 2, Formula 6 leads to the computer result that

$$x_2 = \frac{2.0}{10^5 + 10^5} = 1.0000000 \times 10^{-5},$$

a perfectly rounded root.

The inaccuracy of Example 5 cannot be so simply cured, because it is inherent in working with only 8 decimals, as is revealed by the rapid change of the roots with changes of a, b, c. The best known cure is to identify the delicate part of the computation and use greater precision for it. So Kahan's algorithm uses *double precision* (here 16 significant decimals) in the computation of $u = b^2 - 4ac$, followed by rounding to single precision. The rest of the computation does not need extra precision and is done in the normal way. There is a small penalty in the extra time required for that double precision computation, but it is a negligible part of the total time, spent mostly on scaling and otherwise detecting and correcting overflow or underflow possibilities.

Recomputation of x_1 in Example 5 looks like the following:

$$\begin{aligned}
z &= -b = 4.0000000,\\
y &= b^2 = 16.0000000\ 0000000,\\
w &= 4a = 4.0000000\ 0000000,\\
v &= wc = 15.9999996\ 0000000,\\
u &= y - v = 0.0000004\ 00000000\ 0000000\\
& = 0.0000004\ 0000000, \quad \text{returning to single precision,}\\
t &= \sqrt{u} = 0.0006324\ 5553,\\
s &= z + t = 4.0006325,\\
r &= 2a = 2.0000000,\\
x_1 &= s/r = 2.0003163, \quad \text{rounding up.}
\end{aligned}$$

Note that x_1 is in error by only 0.72 of a unit in the last decimal place.

We shall not discuss scaling and dealing with possible overflow and underflow. The details are many and technical, and they depend intimately on particular features of the computer on which they are carried out. They are extremely important to actual computing but less generally interesting than the ideas just presented. One of the obvious features involves testing whether any or all of a, b, or c are zero.

Conclusion

We have described some of the pitfalls of applying the quadratic formula blindly with an automatic digital computer. We have given sound cures for two of the pitfalls and indicated what other work has been done to create a first-class algorithm for solving a quadratic equation.

The quadratic equation is one of the simplest mathematical entities and is solved almost everywhere in applied mathematics. Its actual use on a computer might be expected to be one of the best understood of computer algorithms. Indeed it is not, and some more complex computations were studied first. The fact that the obvious algorithm is so subject to rounding error is not very widely known by computer users or writers of elementary textbooks on computing methods or, certainly, most writers of mathematics textbooks. Of course, it is familiar to specialists in numerical analysis. Thus, even in this elementary problem, we are working at the frontiers of common computing knowledge.

The majority of practical computations are understood far less well than the quadratic equation. A very great deal of difficult research and development remains to be done before computers will be used as wisely and well as they can be. Almost certainly various parts of the computations for weather forecasting contain pitfalls like those of the quadratic equation, and ignorance of these pitfalls is introducing computational errors that are interfering with progress in weather forecasting. The same can be said about most other nontrivial fields of scientific computation.

The moral of the story is that users of computers for mathematical problems require some knowledge of numerical mathematics. It is not sufficient to learn some programming language and then simply translate formulas from a textbook of pure mathematics into the language of a computer. The formulas and algorithms to be found in most mathematics texts were devised for the exact arithmetic of the real number system. In order to understand the behavior of the formulas when they are used with the approximate arithmetic of computers, the would-be scientific computer must consult people and writings specifically concerned with machine computation.

A further moral is that teachers of arithmetic and algebra should consider the presentation of their subject in the light of its practical application. The quadratic formula is very important for understanding quadratic equations and should be taught. But why hide the fact that it will not stand up to practical application on computers? In days gone by, when few quadratic equations were actually solved, it hardly mattered. But today, when quadratic equations are solved wholesale inside computers, textbook writers should also present material that is not only theoretically correct and easily taught but also easily transferable to practical use. This is a real challenge. In the last analysis, the challenge will be met by persons who have rich backgrounds in both mathematical theory and practical application. This is perhaps the most convincing reason for maintaining the mathematical sciences as a unity.

BIBLIOGRAPHY

L. F. Richardson, *Weather Prediction by Numerical Process* (1922); Reprinted by Dover Publications, New York, 1965.

I. A. Stegun and M. Abramowitz, "Pitfalls in Computation," *SIAM Journal*, *4*, 1956, 207–219.

R. W. Hamming, *Numerical Methods for Scientists and Engineers*, McGraw-Hill, New York, 1962.

E. L. Stiefel, *An Introduction to Numerical Mathematics*, Academic Press, New York, 1963.

BIOGRAPHICAL NOTE

George E. Forsythe heads the Computer Science Department at Stanford University. When he founded it in 1961, it was one of the first such departments in a United States university. From then until 1965, he also directed Stanford's Computation Center. Forsythe served as president of the Association for Computing Machinery. His interest in computing began early, when, as a seventh-grader, he tried using a hand-cranked desk calculator to find the decimal expansion of 10,000/7699. He wanted to see how the digits repeated. In World War II, he encountered more important computation problems as an Army Air Force meteorology officer. He first formally entered the then-nascent field of computer sciences while associated with the Institute for Numerical Analysis of the National Bureau of Standards, Los Angeles.

Among the most conspicuous trends in modern mathematics is the upsurge of "abstract" algebra. Almost every mathematical theory today has an algebraic facet. The structures with which modern algebra is concerned have been compared to the grin of the Cheshire Cat in *Alice in Wonderland*, which remained visible after the cat itself faded away. Algebra's power to generalize often leads to great economies. By emphasizing how seemingly different problems are basically alike, it suggests how the solution of one problem can be adapted to help solve another.

The Algebraization of Mathematics

Samuel Eilenberg

The modern algebra that is increasingly permeating other parts of mathematics is concerned with abstract structures. For a typical example, consider the ordinary plane, which we may think of as the surface of a very large table. Let us agree that a "move" consists of shifting the table and also, if we wish, turning it around its center as pivot. Two such moves may be combined by carrying out one after the other. If we first perform move M and then move N, the resulting combined move is called the *composition* MN of moves M and N. Composition of moves has interesting formal properties. There exists a move I, called the identity or neutral move, which consists in not displacing the table at all. For this move I, we have $MI = M = IM$. For each move M there also is a move M^{-1}, the inverse of M, that undoes what M did, that is, is such that $MM^{-1} = I$. There is also the so-called associative rule for composition: $M(NP) = (MN)P$.

These formal properties of moves occur also in other situations, and, by the process of abstraction from these various examples, algebraists have arrived at the notion of a *group*. A group is any set of elements with a composition law that has the formal properties described above. Thus, the moves described above form a group, but there are also many other groups which may occur under completely different circumstances.

Here are some examples. The integers (the positive and negative whole numbers and 0) form a group if the composition is ordinary addition. The positive fractions form a group if the composition is ordinary multiplication. In the first case, the neutral element is 0; in the second case, the neutral element is 1. In the three examples considered thus far, the group contains infinitely many elements. We give next an example of a finite group, one containing just six elements. Consider the three numbers, 1, 2, 3 and all ways in which we can order, or permute, them. There are six of these permutations:

$$(1,2,3),\quad (1,3,2),\quad (2,1,3),\quad (3,1,2),\quad (2,3,1),\quad (3,2,1).$$

We may think of each permutation as an instruction for rearranging three objects. For instance, (2,3,1) is the instruction: Put the second object in the first place, the third object in the second place, and the first object in the third place. We can now define *composition* by requiring that, for instance, the composite

$$(2,3,1)(3,2,1)$$

be the permutation consisting of carrying out, first, the permutation (2,3,1) and, then, the permutation (3,2,1). Thus, we have

$$(2,3,1)(3,2,1) = (1,3,2).$$

We easily see that in this group the element (1,2,3) is a neutral element.

Let us now give names to the six elements of our group, for instance,

$$a = (1,2,3),\qquad b = (1,3,2),\qquad c = (2,1,3),$$
$$d = (3,1,2),\qquad e = (2,3,1),\qquad f = (3,2,1).$$

The composition rule stated above then leads to the "multiplication table."

Table 1

	a	b	c	d	e	f
a	a	b	c	d	e	f
b	b	a	d	c	f	e
c	c	e	a	f	b	d
d	d	f	b	e	a	c
e	e	c	f	a	d	b
f	f	d	e	b	c	a

In doing group theory, it is sometimes important not to think of the elements of the group just considered as permutations of three objects but simply as six elements subject to the composition rule expressed by our multiplication table. This is what algebraists mean when they say, "We consider this group as an *abstract* group." Every group with six elements subject to the same multiplication table is called *isomorphic* to the group of permutations of three objects, and algebraists consider isomorphic groups as essentially identical.

We note here an example of the general mental process of abstraction. In a similar way, we can consider the group of all permutations of n objects where n is any integer. These groups are called the symmetric groups and are the simplest examples of finite groups. There are many others.

It might seem at first glance that finite groups would be the simplest groups to study, but this is not so. As a matter of fact, some of the most difficult unsolved problems in algebra are connected with the theory of finite groups. For instance, an apparently simple question is: "How many groups are there consisting of exactly n elements?" We mean, of course, "How many *nonisomorphic* groups of n elements are there?" For some values of n, the answer is very simple. If $n = 5$, for instance, there is only one group with n elements. This group may be thought of as a group of rotations of a plane consisting of the neutral element (rotation of 0°), a rotation of 72°, a rotation of 144°, a rotation of 216°, and a rotation of 288°, all in the same direction. The same is true for every n which is a prime, that is, not a product of two integers other than 1. When n is not a prime, however, the problem of describing the totality of groups with n elements is still not completely solved. A few years ago most algebraists felt that the solution of this problem would not be achieved in our lifetime. Today, after an important breakthrough in the theory of finite groups achieved by John Thompson at the University of Chicago, the mood of the expert is much more hopeful.

The origin of group theory goes back to a classical problem in algebra, that of solving algebraic equations. Even the ancient Babylonians knew how to solve certain quadratic equations, and, during the Renaissance, Italian algebraists discovered formulas for solving equations of order 3 and 4, such as

$$x^3 - 2x^2 + 5x + 10 = 0 \qquad \text{or} \qquad x^4 + 7x^3 + x^2 - 6x - 2 = 0.$$

The search for a formula for solving an equation of order 5 or higher was unsuccessful, and in the beginning of the nineteenth century a young Norwegian genius, Abel, proved that such a formula does not exist. A little later, an even younger Frenchman, Galois (he was killed in a duel at the age of 21) found a method of deciding which equations can be solved by formulas involving radicals. It turned out that the key to this problem was the concept of a group and the study of certain properties of groups.

Group theory developed into an important part of algebra. Its applications reach into most of the other branches of mathematics as well as into such neighboring fields as crystallography and theoretical physics. The singling out of the notion of group has been an important turning point in the development of algebra.

Almost equally important notions of *ring* and *field* have evolved out of the study of numbers. For numbers we have two compositions, namely addition and multiplication. The observation of their formal properties has led to singling out the abstract notions of ring and field. A ring is a collection of numbers or other elements that can be multiplied and added the same as ordinary numbers but without the requirement that division by nonzero elements be always possible. For instance, the integers (positive, negative, and zero) form a ring. A field is a collection (set) of numbers or, rather, objects that can be added and multiplied so that all rules of ordinary arithmetic are preserved except perhaps the rules involving the concept of one number's being greater than another. For instance, the rational numbers (positive and negative fractions and 0) form a field, as do the real numbers (rational and irrational numbers) and the so-called complex numbers. Another example of a field is the collection of all numbers of the form $a + b\sqrt{2}$, where a and b are fractions. There also exist fields containing only finitely many elements. For instance, let us define addition and multiplication for the three numbers 0, 1, 2 not in the usual way but according to the following tables. It turns out that

Table 2

+	0	1	2
0	0	1	2
1	1	2	0
2	2	0	1

×	0	1	2
0	0	0	0
1	0	1	2
2	0	2	1

this definition gives a field consisting of exactly three elements. The discovery of such finite fields is also due to Galois. The importance of finite fields for the rest of mathematics has been increasing at a surprising rate during the last decades.

Traditionally, mathematics has been divided into three large branches: geometry, algebra, and analysis. Both algebra and analysis are concerned with numbers, functions, and equations. The chief distinction between them is that algebra was supposed to be concerned with processes that are essentially finite while analysis was to include all that involved infinite processes. With time, geometry has eroded, part of it erupting into algebra and part flowing into analysis. While these processes were taking place, a new branch of geometry called topology came into being. It established strong contacts with algebra (in fact a huge part of topology is called *algebraic topology*) as well as with analysis. The borderlines of the various fields were never too clear, and they continued to shift as a result of new discoveries. One thing has, however, become clear and fairly universally accepted: Every branch of mathematics, whatever it may be, has an "algebraic part" to it. The pages that follow will try to cast some light on this phenomenon.

By way of an example, let us consider a railroad company. The simplest railroad company is one that provides service between two points, say A and B. It has a single track running from A to B and a single train that makes the trip from A to B, stopping long enough to unload and reload, then returning to A, etc. Now compare this with the operation of a huge railway network, with its complicated schedules, rates, and signaling and switching equipment. Basically, the two have the same objectives, to transport people and goods on cars pulled by locomotives on rails. The additional complications resulting from size and complexity may justly be called the "algebra of railroading." Similarly, a rudimentary telephone company would have two subscribers and a single wire joining them. It is very likely that some of the problems of a large railway network will resemble some of the problems of a large telephone company and also, perhaps, some of the problems encountered in traffic control in a metropolitan area. It is also possible that the same mathematical methods used to solve the problems in one case will apply to the others as well. The mathematics or algebra involved may be already in existence, in which case we shall speak of a new application of such-and-such an algebraic theory. More likely, though, the necessary algebra will have to be developed for the purpose, in which case a new subfield of algebra, which might be called the "algebra of flows," will have come into being. Of course, it might turn out that the finite methods of algebra will not suffice to answer the questions and that analysis and differential equations will have to be called on to help.

The penetration of algebra into other fields is witnessed to the greatest extent in mathematics itself. One hears increasingly often the statement that mathematics is becoming more and more algebraized. The relation of algebra to other parts of mathematics has been compared with the relation of mathematics to other sciences. In the mathematical theory of fluid dynamics, one represents an exceedingly complicated and many-sided phenomenon, the motion of a fluid (for instance, the flow of air past an airplane wing) with a mathematical model that consists of a system of functions obeying certain partial differential equations (the precise meaning of these terms need not concern the reader at the moment). The model disregards a large number of important properties of fluids. It is a drastically simplified and ridiculously skimpy picture of reality. But because it is so oversimplified, we can treat it by mathematical methods, and because it mirrors correctly some physical reality, it permits us to draw important conclusions. Similarly, the algebraization of a part of mathematics involves a drastic oversimplification of the mathematical concepts we study, but by concentrating on only some aspects of mathematical reality, we are able to use our knowledge of algebra to study other parts of mathematics.

As an example, consider a spherical container filled with a liquid. Suppose that the liquid is subjected to a smooth motion and that, as a result of this motion, each particle of liquid finds itself in a new position. Then a basic theorem of topology, called the "fixed-point theorem," asserts that at least one particle of liquid will end up in its original position.[1] We can now consider other shapes of containers and ask how many particles will have to end up in their original position. Here, to answer this question, we need algebraic topology. Without going into detail, which would require many pages of preparation, we say only that the shape of the container and the nature of the motion of the particles of liquid can be characterized by a finite set of numbers and that, using these numbers, we can compute the number of particles that end up in their original position. The formula that does this is known as the "Lefschetz fixed-point formula," given by S. Lefschetz in Princeton. Some exciting recent work concerns extensions of the Lefschetz fixed-point formula, which tie it to the theory of differential equations.

Thus far, we have described algebra as, essentially, a part of mathematics used by other mathematical fields as well as by applications outside of mathematics. This is a one-sided picture. Algebra in its many ramifications is a large and old field, and as such it poses its own problems and provides its own inspiration. Many professional algebraists work within established areas of algebra. Others, like this writer, have come to algebra from other fields, forced by sheer need. They treat algebra as a convenient framework and language in terms of which other mathematics can be better understood. If in this process new algebraic methods have to be found (and that is usually the case), then so much the better. One then witnesses the creation of a new subfield called "such-and-such algebra." For example, a new field called "homological algebra" was created in precisely this way.

Closely related to homological algebra is the theory of *categories* and *functors*. A category A has "objects" A, B, C, and so on, and arrows

$$A \xrightarrow{f} B, \qquad C \xrightarrow{h} D,$$

and so on. Two consecutive arrows,

$$A \xrightarrow{f} B \xrightarrow{g} C,$$

[1] To prevent misunderstandings, it may be worthwhile to state that, in talking about a liquid moving in a container, we do not really have in mind a physical problem; the physical imagery is only shorthand to avoid long-winded mathematical discussions.

may be composed to give

$$A \overset{gf}{\rightarrow} C.$$

This composition is associative. Each object A has an *identity*, that is, an arrow

$$A \overset{1\Delta}{\rightarrow} A$$

which, when composed with any other arrow, does not change it. Functors are simply ways of transforming one category into another. There are many examples of categories and functors in various mathematical disciplines.[2] The theory of categories and functors is really a way of expressing and investigating certain common features of mathematical structures.

The basic definitions and facts of this theory were so naïve that to some it seemed highly unlikely that such a theory could have any impact. Nonetheless, only twenty-five years after the definitions were laid down, it has become clear that the theory is appearing with ever-increasing frequency in many developments of modern mathematics. Sometimes it is helpful only in organizing the formalism, but frequently the contribution is more substantial. One recent application of the theory of categories may be witnessed in the theory of automata, computer languages, and abstract linguistics. There are good reasons to believe that, as a result of research still in progress, these theories will be drastically altered by the use of the theory of categories.

BIBLIOGRAPHY

S. MacLane and G. Birkhoff, *A Survey of Modern Algebra*, Macmillan, New York, 1965.

BIOGRAPHICAL NOTE

Samuel Eilenberg was born in Warsaw, Poland, and received his university training there. He came to the United States in 1939 and started his teaching career at the University of Michigan. Since 1947 he has been a professor at Columbia

[2] For those familiar with the terms, we list some examples. The category of groups: here the objects are groups, and the arrows (technically called morphisms) are homomorphisms of groups. Category of topological spaces: the objects are topological spaces and the morphisms continuous mappings. Category of differentiable manifolds: the morphisms are differentiable mappings. Category of vector spaces: the morphisms are linear transformations. Now some examples of functors. The rule which associates with each topological space its one-dimensional homology group and with each continuous mapping of one space into another the induced homomorphism of homology groups is a functor from the category of topological spaces to that of Abelian groups. The rule which associates with each differentiable manifold the vector space of differentiable functions defined on it and with each differentiable mapping the induced linear mapping of the vector space is a functor from the category of differentiable manifolds to that of vector spaces.

University. His research started in point–set topology at the Polish School. He gradually shifted to algebraic topology and from there to algebra, sometimes with and sometimes without a topological flavor. He is at present interested in some of the algebraic aspects of computation: automata and languages. Besides having mathematical interests, Professor Eilenberg is a collector and connoisseur of Indian and Southeast Asian art.

Economic analysis has, in the last twenty years, become predominantly mathematical. This is particularly true in the United States, where doctoral candidates now substitute various courses in mathematics for at least some of the traditional foreign language requirement. Economic problems involving optimum decisions by government and business or stable growth of an economy have analogies in problems of physics and engineering that have long been successfully treated mathematically. But economics has outgrown the days when it merely aped the physical sciences in applying mathematics. The author suggests that in the coming era economics may call forth its own branch of mathematics or provide inspiration for great new mathematical discoveries.

The Role of Mathematics in Economics

Lawrence R. Klein

Economics is a mathematical discipline. This assertion may seem strange to the traditional political economist, but mathematical methods were introduced at an early stage (Cournot, 1838) in the two-hundred-year history of our subject and have been steadily growing in significance. At the present time and, essentially, since the end of World War II, mathematical methods have become predominant in American economics. The mathematical approach was originally inspired in Europe and England but it has flowered in America, with no little stimulus from European immigrants. The mathematical approach is steadily gaining favor throughout the world, especially because the younger generation in developing economies is embracing the new methods and because the socialist countries have shed a previous bias against the use of mathematical methods in economics. It is clear that the future development of economics will see continued and increasing use of mathematics, although it would be rash to assume that the future course of economic analysis will be as predominantly mathematical as it has been in the last twenty years.

The Economic Problem

A favored definition of economics (Lionel Robbins, 1932) is ". . . the science which studies human behavior as a relationship between ends and scarce means which have alternative uses." Whether or not we accept this definition as bracketing all of economics, it is a good starting point for our discussion of the role of mathematics. I might want to sharpen this definition by noting that economists try to select among alternative uses of scarce resources in such a way as to make the most efficient (or least wasteful) employment of resources to achieve stated ends.

Stated in this way, we see clearly that economics involves *optimization*, and this is the engine that produces principles of economic analysis. We have either a maximum problem or a minimum problem, which is a compelling reason for the use of mathematics. An abstract economy is viewed as consisting of numerous consuming and producing units, who make optimal decisions about their own economic behavior, given market prices, and then interact with one another to clear supply and demand in markets to determine prices.

Economic theory usually begins with an analysis of the individual consumer who attempts to *maximize* his satisfaction, subject to a budget constraint (or to minimize budget outlays for the attainment of any given level of satisfaction). The theory then takes up the analysis of producers who strive to maximize profits, subject to a technological constraint (or minimize cost for reaching a given output level, subject to a technological constraint). These are the typical optimization problems of economics.

The standard mathematical formulations of these problems are as follows. The consumer problem is to maximize a utility function

$$u(x_1, \ldots, x_n)$$

of quantities $x_1, \ldots, x_n$ of goods and services consumed, subject to the requirement of living within a fixed income

$$y = p_1x_1 + \cdots + p_nx_n,$$

where $p_1, \ldots, p_n$ are the respective prices of the goods and services consumed. The producer problem is to maximize income minus production costs

$$p_1x_1 + \cdots + p_nx_n - (w_1l_1 + \cdots + w_ml_m),$$

where $l_1, \ldots, l_m$ are inputs of factors of production and $w_1, \ldots, w_m$ are their costs, subject to the restraint imposed by a technical production function

$$f(x_1, \ldots, x_n, l_1, \ldots, l_m) = 0$$

of the quantities $x_1, \ldots, x_n$ of goods and services produced and of the production factor inputs $l_1, \ldots, l_m$.

These two formulations pose the economic problem as the maximization of utility (satisfaction), subject to a budget constraint, and the maximization of profit, subject to a technological constraint. We could also formulate minimum problems that seek minimum production costs for producing a given combination of outputs and the least-cost budget to achieve a given level of utility.

Treatment of Optimization Problems

The consequences of these maximization or minimization problems have been enormous for economics in building a set of rules of behavior. Nearly all economic truths have some root in these or closely related propositions. The original mathematical attack was quite straightforward. Assume that u and f are smooth continuous functions (with first and second derivatives), and optimize according to the rules of the differential calculus, given market prices. The necessary and sufficient conditions for optimization define the well-known demand and supply functions of economics and establish many properties of these functions.

For the problem as I have stated it, these solutions are well established and have been in the literature of economics for more than fifty years. Refined points are made from time to time but the ramifications of this theory were made clear in mathematical treatments by Pareto (1896), Slutsky (1915), Fisher (1892), Hotelling (1932), Frisch (1932), Hicks and Allen (1934), Samuelson (1947).

In the 1930's, and again after World War II, these problems received extended mathematical treatment. The extensions were to optimize over time either continuously or in finite incremental periods and to enlarge the number of side conditions. In stochastic models (i.e., those that incorporate chance), uncertainty about future conditions such as price can be introduced. Also, we can allow for the accumulation of tiny neglected factors that always influence human decisions.

The subjective nature of the utility function, $u(x_1, x_2, \ldots, x_n)$ led to analysis of conditions in which the results of optimization would be invariant under transformations of the function and to study of the possibility of deriving a utility function, starting from objective demand functions. The latter problem became known as the integrability problem.

It may be remarked that the early development of mathematical economics followed the steps of physics and engineering. There are many analogies between the classical methods of mathematical economics and the laws of mechanics, thermodynamics, and similar branches of science. In some cases, there was a tendency to draw strict analogies that could hardly be rationalized in terms of economic behavior.

An idea that received much encouragement from J. Von Neumann was that mathematical economics should draw upon different branches of mathematics that were more suited to the peculiar nature of the economic problem and economic variables. It was even suggested that new mathematical methods might be developed that would be tailored to economics. In the sense that mathematicians of the eighteenth and nineteenth centuries developed methods that were suited to the problems of physics, we might hope that modern mathematicians would receive inspiration from the problems of economics, and social sciences generally. To some extent, this development has occurred in linear programming and optimization theory for situations in which the ordinary methods of differential calculus do not apply. It is up to the mathematicians themselves, however, to decide the significance of this line of development in modern mathematics.

Modern Treatment of Theories of Consumption and Production

The subjective nature of utility has led to new approaches. In their book on the *Theory of Games and Economic Behavior*, Von Neumann and Morgenstern (1944) give axioms of choice. We now find much work on the algebra of choice. Axiom systems are developed in a search for the simplest set that will establish a theory of consumer choice. If these axioms are made stochastic or the choice decisions are made stochastic, the problem then involves the calculus of probability as well. This part seemed incidental to the main argument but has turned out to have a bigger impact on mathematical economics than the strategy of play.

The axiomatic approach is, in a way, more fundamental. It goes back to first principles on which we can all agree and develops laws of consumer behavior. The more direct utility approach assumes the existence of a mathematically tractable utility function and derives the same laws of consumer behavior from its maximization.

The theory of the productive firm as set forth in accepted economic theory, called the neoclassical theory, has been criticized as being unrealistic and not representative of the thought and decision process of modern businessmen or engineers. The nonmathematical versions of this theory have been attacked as much as the mathematical formulations, but the latter have been especially criticized by the cryptic observation that most business decision makers do not know how to form a partial derivative. The first-order maximization conditions, defining in their solution optimal rules of behavior, are not accepted as relevant.

The theory of linear programming that developed in the United States during World War II (in the U.S.S.R. as early as 1939) is regarded as a mathematical model that is more realistically stated in terms of actual business planning and decision making. Its original development in this country was associated with military decision making, but the Soviet examples cited by Kantorovich (1939) dealt, at an early stage, with production economics.

Typical programming problems are associated with blending of gasoline (refinery production) and the preparation of animal feed mixes. In the latter problem, we seek minimum cost diets that supply at least enough materials of specified types for nutritional requirements. It is obvious how the calculations for this problem contribute to the efficient running of livestock farms or the production of feed mixes for animals. These are typical problems that arise in the theory of the productive firm.

A linear programming model may be derived from the analysis of production processes subject to capacity constraints. If each process is capable of turning out a given point in the product space, linear combinations (or averages) of processes determine planes that form the boundaries of production possibilities constrained by capacity limits. These boundary constraints, together with the linear combination of processes, give a set of linear inequality constraints. Total receipts are then maximized, subject to these constraints.

Corresponding to this constrained maximum problem, there is an associated minimum problem. The constraints of the primary problem become the weights in the criterion function of the dual. Correspondingly, the weights in the criterion function of the primary problem are constraints for the dual. Also, row elements in the matrix of primary constraint inequalities become column elements in the dual constraint inequalities. The dual theory can be used in the theory of the firm to express the problem of finding minimum cost points for producing given levels of output. Thus, the theory can be approached from two viewpoints — profit maximization and cost minimization. The first corresponds to the primary problem and the second to the dual. The fact that a linear program and its dual can be reduced to a two-person zero-sum game brings back the relevance of game theory to the analysis of economic production, but this was not the original motivation for the application of game theory to economics. The dual theory of linear programming is, however, a modern representation of the traditional theory just formulated, in which the standard theory of the firm was stated both as a maximum and a minimum problem.

An added dimension occurs in optimization problems if we deal with optimization over time. In some problems where derivatives or integrals are involved, such as in capital theory (capital stock is an integral of past investment) or in inventory theory (inventory speculation depends on price change — derivatives), the optimization procedures lead one into the calculus of variation.

Equilibrium: Existence and Stability

While there is great merit in recognizing that to economize is to optimize (to find either a least-cost or maximum-gain solution to an economic problem), this principle is not the only fountainhead of theorems in economics or source of application of mathematics to economics. There is another great branch of economics that deals with working of markets or finding prices that clear markets.

Walras (1874) made a great step forward when he formulated the problem of general equilibrium in economics as the finding of a solution to a set of simultaneous equations. Optimal consumer behavior generates a set of commodity demand and factor supply equations. At the same time, producer behavior generates a set of commodity supply and factor demand equations (see Table 1).

Table 1

Households	Productive Firms
Demand for goods and services	Supply of goods and services
Supply of labor services	Demand for labor services

Households and firms are not paired off in any special way, but the totality of households and the totality of firms generate *market* demand and supply functions. If we say that there are together n commodities and factors, we have a set of n market equations whose solution will be equilibrium prices.

This looks like a problem of descriptive economics, showing how prices might be found to clear markets in equilibrium, but it, too, may be given an optimal interpretation in relation to welfare economics. There is a sense in which the prices that satisfy this system of equations lead to an optimal allocation of resources in economic society. This is the idea of Pareto, who stated that optimality exists if a state is reached such that no person in society can be made better-off without making someone else worse-off. It is then a problem in welfare economics to show that the prices that clear markets (in a competitive society) are actually optimal in the Paretian sense.

It was a great step forward on the part of Walras to conceive of the economy as a system of simultaneous equations, but he contented himself with a counting of equations and variables, and we later learned that this was far from adequate. Many economists recognized the mathematical problem in showing the existence of equilibrium, but it was Abraham Wald who first reported a rigorous statement for the existence of such solutions to the problem of general equilibrium. There has since been much improvement in the statement of this problem, making use of linear programming theory (solution to a linear model), properties of convex sets, and fixed-point theorems.

If equilibrium can be shown to exist, what are the conditions under which it is stable? The "law of supply and demand" is given explicit mathematical formulation to handle this problem. It is assumed that a competitive society acts in the following way: If there is an excess of supply (demand) in any market, price falls (rises) in order to clear that market. This reaction is not based on optimality theory but is thought to describe the workings of competitive markets. The corresponding mathematical formulation is a system of differential equations. Conditions under which the solutions to such systems of differential equations are stable have been derived.

It is in the analysis of equilibrium and its stability that we might see much work on mathematical economics in the planned societies of the socialist countries. Although the solution to the market equations of a competitive economy may be stable, they may oscillate. This has long been recognized as a source of the economic phenomenon known as the *business cycle*. If the actual economy in real life is not perfectly competitive, we may find oscillations that are not stable; moreover, it is not an established theorem that competitive oscillation is stable. Stability has been established only in special cases. Finally, it must be recognized that movement of prices may occur under random or nondeterministic circumstances. If the differential equation system is nonlinear, we are not in a position to say, in general, what its stability properties will be under random shock.

It is for reasons of instability that socialist economists have been wary of using the market mechanism for solving the problem of allocating scarce resources to achieve given ends. They may simulate markets or seek computer-based solutions of the equilibrium equations to obtain prices that are free of oscillations and possible unstable movements. A socialist planner must, however, conceive of the economy as a set of Walrasian equations and seek dynamic solutions that bring him to his goals. This will be one of their mathematical problems.

Macroeconomics

With compact matrix and vector notation, a complicated mathematical model of economic life, even at the general equilibrium level, can be written in a deceptively simple form. We may forget that these compact expressions deal with millions of households, firms, commodities, and prices. To obtain expressions that are simple enough for the human mind to comprehend, we have to consolidate. This is known as the problem of aggregation, or the index number problem, in economics. We must make use of a theory of summarization that provides a mathematical bridge between micro- and macroeconomics. An index number of prices, production, wage rates, or similar economic magnitudes is used to stand in the place of all the detailed components.

It is evident that conditions for the existence of aggregative relationships, in an exact form, will be highly restrictive. Some sufficient conditions can be established. For example, Hicks has proved the composite good theorem of consumer behavior, which states that a group of goods whose prices all change in the same proportion can be treated as a single good in deriving the mathematical laws of market demand. But prices do not conveniently change this way, and other sufficient conditions are equally restrictive. We, therefore, must accept the equations of macroeconomics as approximate. There is an error of aggregation. We generally build macroeconomic systems of equations by analogy with those of individual theory. In deriving propositions from such systems, we must keep in mind that they are not exact, and a way of doing this might be to treat them as stochastic equations. This introduces a new dimension into the analysis of their properties.

The study of the business cycle in dynamic economics is traditionally based on the equation systems of macroeconomics. One such simple model of the economy as a whole, due to N. Kaldor (1940), is

$$S = f(Y), \qquad \text{aggregate savings function,}$$

$$I = g(Y, K), \qquad \text{aggregate investment function,}$$

$$\frac{dY}{dt} = h(I - S), \qquad \text{income adjustment,}$$

$$\frac{dK}{dt} = I, \qquad \text{relation of capital to investment,}$$

$$S = \text{savings}, \qquad I = \text{investment},$$

$$Y = \text{income}, \qquad K = \text{capital}.$$

This simple aggregative model resembles dynamic theories of price determination, cited above, in studying the stability of general equilibrium. By aggregation, it has reduced the study of the economy as a whole to a few relationships associating a small number of aggregative variables. Although it appears to be simple, it is powerful in revealing much of the business cycle process of the economy. The original formulation of this model was put forward by Kaldor as a dynamic generalization of the static theories of J. M. Keynes (1935), and Kaldor showed that, if f and g have some plausible nonlinear shapes and if h is a positive function of $I - S$, with

$$0 = h(0),$$

this model is capable of producing a self-maintained oscillation resembling the business cycle.

Models of this type have become considerably more elaborate with wage rates, price levels, separate producing sectors, monetary relationships, foreign trade, public policy variables, and more elements of economic growth. In fact, the latter subject has caught the fancy of economic theorists, who have turned their attention away from problems of business cycles (assuming them to have been conquered in real life) to mathematical models of growth in mature economies and of development in the emerging economies.

Two major issues appear to have governed the mathematical analysis of growth. A celebrated paper by Von Neumann dealt with a general model of the expanding economy containing many interrelated departments or sectors. He derived conditions for such an economy to have a "balanced growth path," with each sector expanding at the same rate and this common rate equaling the rate of interest. This general result stimulated much research in the construction of two- and three-sector aggregative models that have the property of balanced growth.

In the course of this work, much doubt was cast on the time-honored concept of total capital, measured as

$$K(t) = K(0) + \int_0^t I(\tau)\, d\tau$$

or in discrete terms as

$$K_t = K_0 + \sum_{i=1}^{t} I_i,$$

where K = stock of capital and I = net investment. The cumulation of investment over time was criticized for not allowing for changing differences in the quality of production goods as a result of technical progress. R. Solow conceived the idea of vintage measures of capital as

$$K(t) = K(0) + \int_0^t e^{-\rho(t-\theta)}\, I(\theta)\, d\theta$$

or

$$K_t = K_0 + \sum_{i=1}^{t} \lambda^{t-i} I_i, \qquad |\lambda| < 1.$$

In these measures I_i is the *gross* amount of new producer goods required during the ith interval and λ^{t-i} is a weight factor that accounts for technical obsolescence and physical depreciation. Properties of growth models that were previously established under the older definitions of capital had to be reworked in terms of vintage measures.

The subject of macroeconomics and indexing has been discussed here in only one aspect, that of constructing and analyzing aggregative models of the economy as a whole. Another use is for the question of ideal measurement. Most people are familiar with the concept of "cost of living." Economists have a special way of measuring this concept. They seek an index of the amount of money that a person would have to spend in situation A, as compared with the base situation B, to be as "well off" (achieve the same standard of living) as in B. This leads to formulation of the indirect utility function

$$u[x_1(p_1, \ldots, p_n, y), \ldots, x_n(p_1, \ldots, p_n, y)]$$

and integration of the differential equation defined by

$$du = 0.$$

Solution of this problem shows how the true "cost of living" can be defined in terms of the parameters of individual demand functions

$$D_i = X_i(p_1, \ldots, p_n, y).$$

Other forms of ideal measurement in the field of index construction cover measurement of indexes of output, costs, and productivity. These depend on results in the mathematical theory of the firm the same way that cost-of-living index construction depends on results in the mathematical theory of the household.

Econometrics

While many people use the terms "mathematical economics" and "econometrics" synonymously, we shall draw a distinction. Some people are equally at home in both fields and the membership of the Econometric Society includes both; yet econometricians are concerned mainly with measurement in economics — how to make measurements in theory and how to apply them to realistic situations. In a way, the problems of index construction and capital measurement are econometric problems, but they may be treated as pure theoretical issues in welfare economics or the abstract analysis of growth systems. Measurement problems in econometrics are usually formulated with a view toward realistic application, in principle at least.

One function of econometrics is to estimate the mathematical equations of economic theory. Such estimation usually takes the form of numerical determination of parameter values, within error limits. The demand equations, production relationships, or the equations of macroeconomic models are the subjects of econometric estimation.

The general approach is to recast mathematical economics into a stochastic form. This is quite essential in econometrics, although much of mathematical economics is deterministic. From the law of error distribution, we estimate economic parameters, following established principles of mathematical statistics. The reason that it is absolutely essential to deal with stochastic relationships is that there are errors of aggregation and there are omitted variables. Our simplified mathematical formulations of economic life cannot be sufficiently detailed to embrace all variables that affect economic behavior.

At the simplest level, we are faced with the problem of estimating economic relationships of the form

$$y_t = f(x_t; \alpha) + e_t, \qquad t = 1, 2, \ldots, t,$$

from sample observations. These samples may be collected on the basis of different time points (time-series sample) or different locations (cross-section sample) or a combination of both. They may be observations on individual economic units or on social aggregates. Stochastic error is represented by e_t, and α is a parameter to be estimated. It is a problem in mathematical statistics of some special interest if we note that the samples are nearly always nonexperimental.

More generally, there are many dependent variables in economic relationships and also many relationships simultaneously at work generating our nonexperimental samples of data. T. Haavelmo (1944) did for the econometrics of economic relationships what Walras had done in a previous era for the mathematics of economic relationships. Haavelmo drew the logical consequences for econometrics of the simultaneity in economic life, provided that we cannot generate observational data through experimentation. Since the y_{it} mutually determine each other throughout the system and within each single equation, we cannot estimate a given relationship by itself, using the customary methods of regressing one y_t on all the others and the x_t in that equation. Haavelmo demonstrated that we ought to take account of the system properties and the dependent interrelatedness of the y_t's in estimating all equations of the system. His model would take the general form

$$f_i(y_{1t}), \ldots, y_{nt}; x_{1t}, \ldots, x_{mt}; \alpha_1, \ldots, \alpha_r) = e_{it},$$
$$i = 1, 2, \ldots, n, \quad t = 1, 2, \ldots, T.$$

In this model there are n relationships among n dependent variables (y_{it}) and m independent variables (x_{it}). These relationships are stochastic and are disturbed by n random variables.

There are various methods of parameter estimation that have been proposed and are being used for very large systems, some now consisting of 300 equations.

The objective of econometric research in the estimation of equation systems has been to give empirical content to mathematical economic theory and also to track real world events in the economy as a whole or market subsectors of it. The results of applied econometrics have been used for economic policy formation by national governments, international bodies, businesses, or public organizations. A careful mathematical formulation of economic relationships and their implications, together with mathematical theories of statistical inference, has finally paid off in producing estimated models of economic life that have genuine applicability to important social matters. Mathematics has not been able to do this alone, but mathematics in the service of applied economics with the help of elaborate data collection, computerization, and advances in general economic thinking appears to be the appropriate combination.

All econometrics is not concerned with the estimation of economic relationships by using the methods of statistical inference. An important descriptive problem in econometrics has been the attempt to characterize the distribution of economic quantities. These distribution functions play an important role in estimation of economic relationships, as we have just seen, and they are also of great importance in the whole theory of aggregation and index numbers, taken up in the previous section. Economic welfare obviously depends on the way rewards are distributed throughout economic life. It has, therefore, been a long-standing problem, dating back to the researches of Pareto to explain the person-to-person distributions of income and wealth. It is equally important to be able to characterize the size distribution of firms or the distribution of price changes. The mathematical problem here is to construct a stochastic process, say in the form of a difference or differential equation, that would explain the observed distribution of income and other economic magnitudes. Among the many schemes that have been proposed, a fruitful line of research appears to be the fixing of relationships among transition probabilities that show the percentage of people in income class i in the tth period who move into income class j in the $(t + 1)$th period. D. Champernowne (1953) has shown that, if transition probabilities p_{ij} depend only on the spread between i and j,

$$p_{ij} = q(j - i),$$

when income class limits grow in geometric progression, then the limiting distribution of movement from class to class will be the Pareto type:

$$P(y) = \frac{\alpha}{y_0} \frac{(y_0)^{\alpha+1}}{y} \quad \text{for} \quad y > y_0.$$

By analogous reasoning, it can be shown that if the transition probabilities depend on the ratio of j to i (instead of their difference) and if class limits are in arithmetic progression,

$$p_{ij} = q(j_i),$$

then we have a lognormal distribution of incomes:

$$P(y) = \frac{1}{\sqrt{2\pi}\,\sigma y} \exp\left[-\frac{1}{2\sigma^2} (\log y - \mu)^2\right].$$

These are only two popular parametric forms of income distribution, and much more mathematical analysis of the stochastic generation of distributions in econometrics needs to be made. Other forms have been developed, but this is very much an unsettled problem.

A departure from the view of econometrics as statistical inference is the work done in input–output analysis, a subject developed by W. Leontief. This is economic measurement in great detail and has given rise to extensive mathematical literature. It is mainly nonstochastic, but it could be given a probabilistic interpretation. The purpose of $(I — O)$ analysis in econometrics is to trace intermediate flows among separate producers in a complex, highly interrelated economy. Let us denote the goods shipped from producing sector i to j as x_{ij}. The total output of the ith sector is x_i, and the final demand for its product is F_i. By final demand we mean shipments beyond the intermediate destination of other producers. We mean shipment to final users such as households, government, foreigners. By definition, we have

$$x_i = \sum_{j=1}^{n} x_{ij} + F_i, \qquad i = 1, \ldots, n.$$

The crucial assumption of the theory is

$$a_{ij} = \frac{x_{ij}}{x_j};$$

that is, there is a proportionate relationship between the intermediate input flow x_{ij} and x_j. The technical coefficient a_{ij} is the amount of good i used to produce a unit of good j. By substitution, we have

$$x_i = \sum_{j=1}^{n} a_{ij}x_j + F_i, \qquad i = 1, \ldots, n.$$

In matrix notation, we write

$$(I - A)X = F.$$

It is a tedious job of data collection to determine x_{ij} and F_i, but once it is done for a base year, the main mathematical problem is to evaluate the inverse matrix in

$$X = (I - A)^{-1}F$$

and establish its properties. Knowledge of $(I - A)^{-1}$ enables us to estimate X, the vector of industrial outputs, for different assumed vectors of final demand, F.

This is an indispensable tool in analyzing an economy. Large systems (between 100×100 and 500×500) have been computed and maintained by many nations. It is useful in development planning for emerging countries and in emergency planning for developed countries. It is now being combined with aggregative models, established by inferential methods, and the two together enable us to estimate both the over-all level of economic activity and its industrial composition. Socialist planners are making heavy use of $(I - O)$ results in laying out their n-year plans.

Epilogue

Mathematics has now caught on in economics. A final decision, at the teaching level, on the acceptability of the mathematical approach has been taken in the substitution of mathematics for one or more of the foreign languages traditionally required for the doctorate in the United States. The struggle for recognition is therefore over, and the way mathematical economics and econometrics will develop is a matter of speculation.

It is certain that the level of mathematical competence and sophistication has markedly increased during the past few years. It is also becoming clear that some specific branches of mathematics are to be emphasized; analysis, linear algebra, set theory, topology, probability theory, numerical analysis, and differential equations are outstanding examples. Mathematical economics has outgrown the days of aping physics and engineering in mathematical applications, but there are few examples in which it can be said that our subject has called forth its own branch of mathematics or given inspiration for great new mathematical discoveries. That era may come and in an unexpected way. Meanwhile, the problem is to consolidate the developments that have been made by clearly demonstrating the relevance of the new wave of mathematical methods in economics.

There have been fads in the history of economic analysis. It seems unlikely that the mathematical approach is another passing fad, but, to safeguard the advances that have been made, the coming generation of economists will have to realize that mathematics is no panacea. Economic problems are really very difficult, and the greatest scientific talents that cracked open new discoveries in physics and mathematics have not been able to make as great an impression on economics when they turned their attention to social studies. Perhaps there has been a von Neumann Revolution, as M. Morishima declared in his presidential address to the Econometric Society, but that is an exceptional case in an area of limited scope.

Some branches of economics — economic history, economic philosophy of social goals, parts of political economy — are not likely to lend themselves to mathematical treatment. The economist who is to be an all-around man must master more than the mathematical side of the subject. If the coming generation can do that, and at the same time acquire mathematical tools, they will be in a strong position to obtain powerful new results. Most likely, economic intuition and knowledge of world affairs will suggest the problems to be solved. It will then be up to the new methods of mathematical analysis to work out many of the solutions.

BIBLIOGRAPHY

"Mathematics in Economics," a symposium in *The Review of Economics and Statistics*, *XXXVI*, pp. 359–386, November 1954.

"Mathematics and the Social Sciences," a symposium sponsored by The American Academy of Political and Social Science, Philadelphia, June 1963.

D. Gale, *The Theory of Linear Economic Models*, McGraw-Hill, New York, 1960.

J. Henderson and R. Quandt, *Microeconomic Theory*, McGraw-Hill, New York, 1958.

M. Morishima, *Equilibrium, Stability, and Growth*, Oxford University Press, New York, 1964.

P. A. Samuelson, *Foundations of Economic Analysis*, Harvard University Press, Cambridge, Mass., 1947.

L. R. Klein, *An Introduction to Econometrics*, Prentice-Hall, Englewood Cliffs, N.J., 1962.

A. S. Goldberger and L. R. Klein, *An Econometric Model of the United States, 1929–1952*, North-Holland, Amsterdam, 1955.

W. Leontief, *Input-Output Economics*, Oxford University Press, New York, 1966.

E. Malinvaud, *Statistical Methods of Econometrics*, Rand-McNally, Chicago, Ill., 1966.

BIOGRAPHICAL NOTE

Lawrence R. Klein is Benjamin Franklin Professor of Economics at the University of Pennsylvania, where he is in charge of one of the most comprehensive mathematical models of the U.S. economy. This model, consisting of more than seventy simultaneous equations, is used to analyze the economy and make forecasts for government and business. Professor Klein's dual interest in economics and mathematics began soon after he entered college and continued through his undergraduate days at the University of California at Berkeley despite, as he says, "bad advice from curriculum advisers to tone down my mathematical studies." His was the first Ph.D. degree in economics ever awarded by M.I.T., where he was strongly influenced as a graduate student by Paul A. Samuelson.

Topology, the study of how space is organized, is like the great temples of some religions. That is to say, those uninitiated into its mysteries can view it only from the outside. Nevertheless, even from the spectator's viewpoint, some of the developments of modern topology are startling, and many findings concerning space of many dimensions are directly applicable to physics. While outlining the evolution of topology, particularly in relation to the famous *n*-body problem, the author discusses some of the implications of this towering temple of abstraction. Among recent discoveries is a ten-dimensional surface that is "unsmoothable." If a protruding corner is pushed down at one place, another corner pops up elsewhere. Another new topological theory implies that the solar system is inherently stable: that the planets will continue to orbit as they do today, not for just thousands of years but forever.

The Evolution of Differential Topology

Andrew M. Gleason

It is notoriously difficult to convey a proper impression of the frontiers of mathematics to nonspecialists. Ultimately the difficulty stems from the fact that mathematics is an easier subject than the other sciences. Consequently, many of the important primary problems of the subject — that is, problems which can be understood by an intelligent outsider — have either been solved or carried to a point where an indirect approach is clearly required. The great bulk of pure mathematical research is concerned with secondary, tertiary, or higher-order problems, the very statement of which can hardly be understood until one has mastered a great deal of technical mathematics.

Pure mathematics deals entirely with abstractions. Contrary to popular impression, abstractions are not vague; they can be defined with far greater precision than anything in the real world; consequently, they can support very long chains of logical reasoning. Imagine a biologist who painstakingly studies the digestive process of an amoeba. Suppose, after understanding this to his satisfaction, he retires to his armchair and extrapolates his results through successive levels of the animal kingdom and comes up with a theory of the digestive process in man. It would be absurd to expect that his theory would have any particular relation to the observed facts of human digestion. Yet the mathematician regularly does something rather similar to this. He applies in very complicated contexts theories which were derived in very simple ones, and he not only expects, he finds, that the results are valid, not merely in outline but in detail.

When a mathematician meets a problem he cannot solve, like any other scientist he tries to solve instead some related problem which seems to contain only part of the difficulties of the original. But the mathematician has far more alternatives in choosing a simpler problem than does a chemist or biologist. Other scientists are restricted by nature, whereas the mathematician is restricted only by logical coherence and somewhat vague considerations of taste. Because of this greater freedom, mathematical problems evolve more rapidly, and often the end product seems unrelated to its origins. Yet most of pure mathematics today is concerned with problems that have grown naturally from primary questions of obvious interest, many of which have great practical value. Once one develops a taste for pure mathematics, he is likely to pursue his favorite problems without serious concern for their origins. In fact, it is almost necessary to do so, because no one can keep up with all of mathematics today.

This is all very good for the specialist, but to present these problems to the outside world with no description of how they are related to the broader field is a disservice to mathematics. A person may read, for example, that topology is the study of "rubber-sheet geometry," that is, of properties of geometric figures which withstand stretching. Later he reads that a significant theorem of topology has the consequence that at any given moment there are two points on the surface of the earth which are precise antipodes and which enjoy the same barometric pressure and the same temperature. If he now decides that topology is a hopelessly frivolous subject, I cannot blame him.

I should like to give you a brief look at one of the most famous problems of mathematics, the n-body problem, to sketch how some important questions of topology are related to it and, finally, to tell you about two important recent discoveries in topology whose significance is only beginning to be appreciated.

Differential Equations

To begin with, we must understand something of differential equations.

Suppose the Cleveland police department erected a sign at each street corner in the city which specified that any car arriving at that corner leave along a definite street, no U-turns being permitted. A definite path would thereby be determined for every car. If we knew the traffic directions explicitly, we could compute the path of any car, given its starting point.

An ordinary differential equation presents an analogous problem involving a continuous system of traffic instructions. It specifies a direction at each point of the plane. Imagine a small object moving in the plane and always obeying these traffic directions. It will follow a curve in the plane which has the prescribed direction at each of its points. Such a curve is called an *integral* curve of the differential equation. If the instructions are too hodge-podge, it may be impossible to follow them; however, if we assume that the directions specified at nearby points are nearly parallel, we can prove that there is a unique integral curve through each point of the plane. If we imagine all such curves drawn in the plane, then the plane will be paved with curves. In every small area they will look rather like parallel lines, but in large regions different curves may diverge sharply from one another. To solve a differential equation means to find the integral curves explicitly.

A helpful way to think of a differential equation involves the intuitive notion of infinitesimal segment. The direction at each point may be specified by an arbitrarily short line segment through the point (Figure 1),

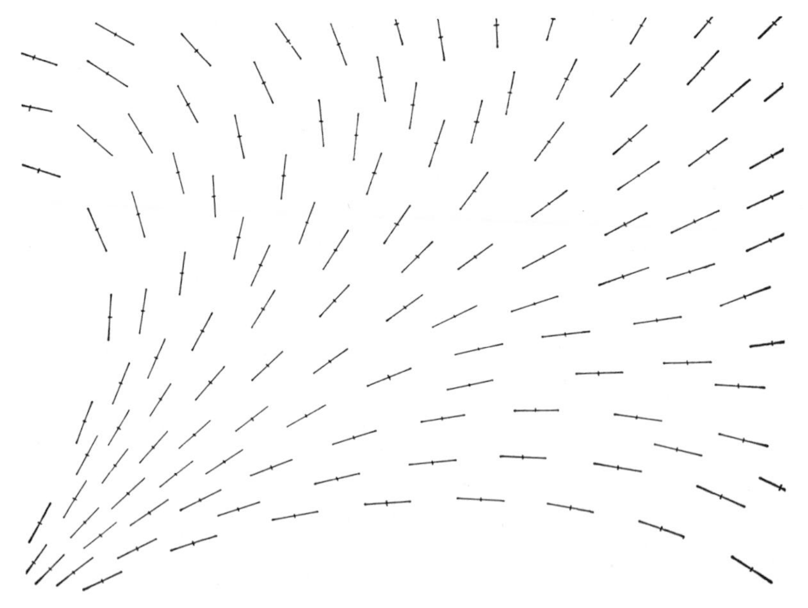

FIGURE 1. *A first-order ordinary differential equation in one variable. Given directions at each point of the plane, find curves which are at each point tangent to the given line segments.*

and we may think of these segments as "infinitely short." These infinitesimal segments are equally infinitesimal segments of the desired curves (Figure 2), since a curved line looks straight when viewed under sufficient magnification. Thus, the differential equation may be regarded as giving the desired curves fragmented into infinitesimal pieces. The process of reassembling these pieces into whole curves is called *integration.*

Returning to the cars moving in the streets, let us suppose that at each street corner, in addition to a direction, a definite speed is prescribed. Then we not only can deduce the path followed by a car, we can also find the exact time required to cover each section of the path.

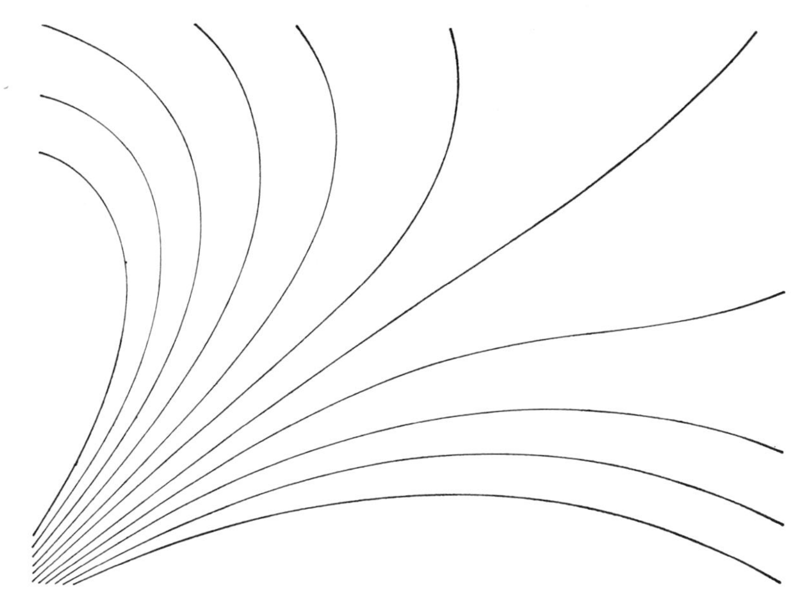

FIGURE 2. *Solution of the differential equation of Figure 1. The plane is covered with curves having the desired direction at each point.*

Correspondingly, in the continuous case we can specify at each point of the plane not only a direction but also a speed. This leads to a new kind of differential equation, in which we assign an arrow at each point of the plane (Figure 3); the direction of the arrow tells us the direction in which an object should move, and the length of the arrow tells us the speed. An object which moves in accordance with these prescriptions must now not only follow a definite curve in the plane, it must traverse each section of the curve in a definite time. To solve this differential equation, we must integrate once to find the paths or integral curves and then integrate again to find the exact times of flight along the paths.

Generalization to Space

We can, of course, generalize this idea to space. A differential equation is the assignment of an arrow at each point of space. An integral curve of the differential equation is a curve which is tangent at each of its points to the arrow prescribed at that point. If the prescribed arrows do not vary too rapidly from point to point either in length or in direction, there will be a unique integral curve through each point of space, and there will be

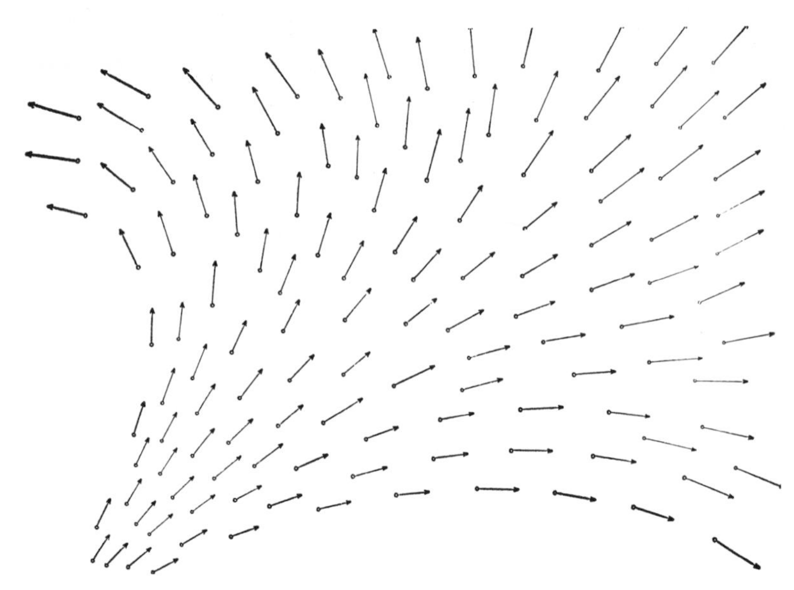

FIGURE 3. *A parametric ordinary differential equation of the first order in two dimensions. A speed (indicated by the lengths of the arrows) is now prescribed at each point of the plane in addition to the direction specified in Figure 1. The integral curves are the same as before, but now times of flight along each curve are also determined.*

a unique motion along each integral curve which conforms to the prescribed speeds. To solve the differential equation means to find specifically the integral curves and the times of flight along them.

Take some curve in space, not an integral curve. The various integral curves emanating from this curve will form a surface (Figure 4). If we start from a system of more or less parallel curves, we will get a system of more or less parallel surfaces. Each of these surfaces will be made up of integral curves. Finding a family of surfaces of this type which pave space nicely is called integrating the differential equation once. With such a family of surfaces explicitly known, we are well along toward solving the problem. Given a point p, we seek the integral curve through p. We know it lies on the unique surface S of our family, which passes through p. Since S is explicitly known, we can confine our attention to S in searching for the integral curve C. Finding C on the (presumably curved) surface S is not very different from finding an integral curve for a differential equation in the plane. The differential equation must be integrated twice more, once to find the curves and again to find the times of flight.

The step to higher dimensions is clear, even if the geometric imagery has only a fictitious significance. In four dimensions we prescribe an "arrow" at each point. Then we try to find a family of three-dimensional

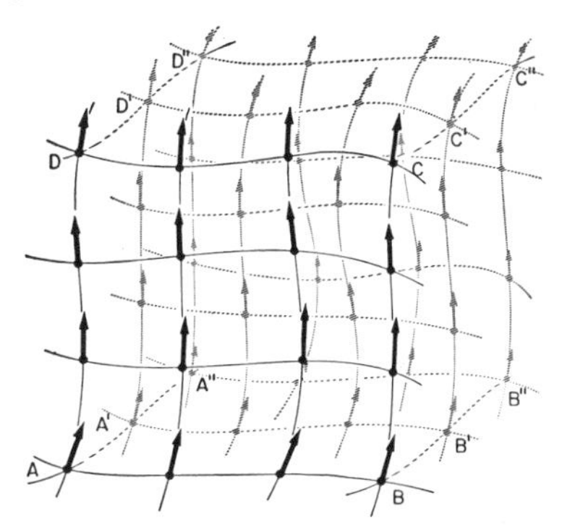

FIGURE 4. *A parametric ordinary differential equation of the first order in three dimensions. Arrows at each point of space determine the integral curves and the time of flight. The integral curves emanating from the nonintegral curve* AB *form a surface* ABCD *called an integral surface. Another curve,* A′B′, *which is more or less parallel to* AB, *generates another integral surface* A′B′C′D′ *more or less parallel to* ABCD. *Surface* A″B″C″D″ *is similarly generated. This leads to a covering of space by a family of integral surfaces. Another family of curves* AA′A″, BB′B″, *and so on, generates another family of integral surfaces. The intersection of two integral surfaces from different families is an integral curve; for example, the integral surface* BB″C″C *meets* ABCD *in the integral curve* BC. [*Courtesy Addison-Wesley, Reading, Mass.*]

surfaces which "pave" the four-dimensional space; that is, we try to integrate the equation once. If successful, we try to integrate again, which means that we try to "pave" each three-dimensional surface with two-dimensional surfaces. A third integration gives us the integral curves; a fourth, the times of flight. In still higher dimensions the situation is the same. We successively seek ways to pave space with surfaces of lower and lower dimension, each of which has the property that it is made up of integral curves. After integrating one time less than the original dimension, we will have found the curves, and one more integration will be required to find the times of flight along the curves.

Planetary Motion

One of the most important applications of differential equations and one of the greatest triumphs of mathematical physics is the Newtonian theory of planetary motion. According to Newton, the *state* of each body in the solar system can be described by six numbers, three to designate its position and three to designate its velocity — that is, the speed and direction of its motion.

If we wish to consider an abbreviated solar system consisting of the sun and one planet, we will require six numbers to describe the state of each body or twelve numbers in all. Thus, the set of all conceivable states of the system can be regarded as points in a twelve-dimensional space. At each instant, the actual state of the system will be a single point in this state-space. As time elapses, the successive states of the system form a curve in state-space; we can say that the actual state moves in state-space.

Newton proposed that the actual state will move in accordance with a differential equation described by his laws of motion and gravity. These laws prescribe at each point of state-space an arrow which tells the velocity the actual state will have if it should pass through this point. (Since the state of a body includes its velocity, a velocity in state-space is what we ordinarily call an acceleration. Newton's laws of motion and gravity describe the acceleration of each body in the system as a function of the state.) The problem of two bodies is thereby reduced to solving a differential equation. To find the motion of a specific system, we must find the integral curve through the starting state and the times of flight along it.

The general problem of two bodies is to find all the integral curves and the times of flight along them. This means we must integrate the differential equation twelve times, once for each dimension of the state-space. This was worked out by Newton, and a remarkable result appeared: The predicted behavior of the planets was in precise agreement with the laws of planetary motion empirically determined by Kepler. As we all know, this theoretical success was one of the great turning points in the history of science.

If we apply the two-body solution separately to the various planets, we are neglecting the interactions between them. The theory also predicts the effect of these interactions. If we take three bodies, the state of the whole system is described by six numbers for each body, so the set of all conceivable states can be regarded as an eighteen-dimensional space. Again, Newton's theory gives us arrows in the state-space which tell us how the states will evolve. To predict the behavior of a specific system we must find the integral curve through the starting state and the motion along it. To solve the three-body problem means to find all the integral curves and the times of flight along them. To do this, we must integrate eighteen times. It is easy to integrate the equation ten times, leaving eight to go. We can interpret this as follows. Given a starting state, we can describe an eight-dimensional surface in the space of states such that all future states will lie on this surface. But an eight-dimensional surface is a long way from a curve. From Newton to Poincaré, many mathematicians tried to find further integrals of the three-body problem without success. Finally Poincaré, in the latter part of the last century, showed that there are no further integrals in closed form [1]. This means that progress can be made only by approximation techniques.

The idea of approximate solutions of the three-body problem, and even of the n-body problem, had occupied the minds of mathematicians from the days of Newton. Many of the greatest mathematicians worked to find efficient methods for predicting the planetary motion. Approximate solutions require a great deal of numerical work, and there were no computing machines, not even desk calculators, in 1800. Nevertheless it was possible to predict, even then, the motions of the heavenly bodies within the accuracy of observations.

The theory has served so well that we accept as commonplace the fact that we can predict the exact circumstances of eclipses for many years to come. But on one occasion, at least, a prediction of the Newtonian theory was dramatic. In the early nineteenth century the observed motion of the planet Uranus (discovered in 1781) was found to be at variance with the Newtonian theory. To resolve the difficulty, Adams and Leverrier independently predicted the existence and approximate location of the hitherto unknown planet Neptune, which was found shortly thereafter within one degree of the predicted spot!

There have been some major theoretical advances in the Newtonian theory in the last few years. In particular I would like to mention the work of Arnol'd. The solar system appears to be stable. We can predict the course of the planets for thousands of years to come, and we find that they do not stray far from the simple Keplerian orbits calculated by neglecting the planetary interactions. But thousands of years is not forever, and it may be that the solar system is unstable. After several billion years, say, the earth might suddenly emerge from the solar system at a speed sufficient to project it irretrievably into outer space. Or two planets might collide — another form of instability. Arnol'd has proved a strong stability theorem concerning Newtonian n-body systems [2]. Slightly overstated for simplicity, his theorem is as follows. If one body is a great deal larger than all the others and if the system starts in a superficially stable state, then it is in fact stable. Of course this result is of mathematical significance only, because it allows for neither relativistic nor quantum theoretic effects and neglects totally the other stars in the universe.

Topological Methods

Let us return to a special differential equation in two dimensions. Suppose that, as in Figure 5, there is an integral curve which winds in-

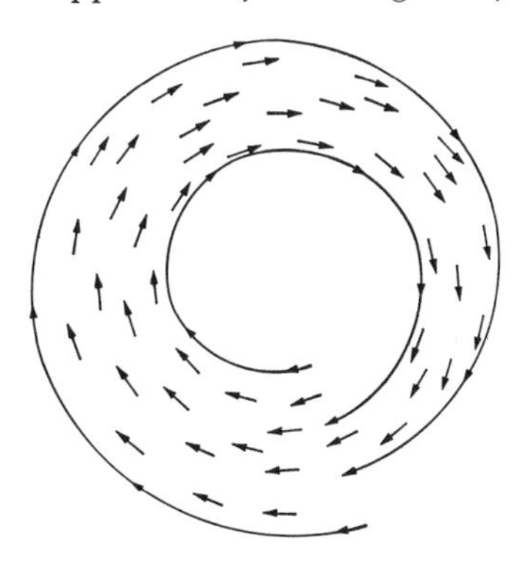

FIGURE 5. *Parametric differential equation in the plane, with spiral integral curves. Because one curve spirals out and another spirals in, there must be an intermediate one which spirals neither in nor out and is therefore a closed curve.* [*Courtesy Addison-Wesley, Reading, Mass.*]

ward and another which spirals outward. It is easy to see, and it can be rigorously proved, that somewhere in between there must be an integral curve which spirals neither in nor out but actually meets itself precisely. Such a curve then continues around and around the same circuit indefinitely. If the differential equation represented a problem in planetary motion, this curve would represent a periodic solution; the system would return at regular intervals to the same state and retrace its motion over and over again.

Note how little we need to know to guarantee the existence of a periodic orbit. In the first place, we don't have to know the inner and outer curves precisely. It is enough to know that one spirals in and the other out. This we may establish through approximate calculation. Secondly, we don't really have to know the differential equation we are dealing with precisely, because a slight change in the equation will leave the same qualitative situation — namely, one integral curve spirals in and the other spirals out.

This is one of the simplest examples of what are called topological arguments. The branch of mathematics which studies them for their own sake is called topology. While some results of topology were known earlier, the subject did not obtain a serious place in mathematics until Poincaré showed that such arguments could prove the existence of periodic solutions in special cases of the three-body problem [1]. Since that time the subject has grown to be one of the most significant branches of mathematics.

Let us look at another topological fact germane to differential equations. We noted that in solving a three-dimensional differential equation one usually first finds a system of surfaces each of which is made up of integral curves. There will usually be many ways to organize the integral curves into surfaces. It might happen that one of these surfaces is a closed surface, that is, a surface which is finite in extent but without edges. Figure 6 shows some closed surfaces: the surface of a ball, known as the two-sphere; the surface of a ring or torus; the surface of a two-holed solid. We can go on to the surfaces of solids with more and more holes. This gives us a sequence of surfaces which are all topologically distinct. Furthermore, every closed surface in three-dimensional space is topologically equivalent to one of the surfaces in this infinite sequence of surfaces.

Let us pause to recall what topological equivalence means. It is usually said that topology imagines the figures to be made of rubber and that two figures are equivalent if one can be stretched or compressed without tearing to look like the other. This is not strictly correct. It is permissible to tear the figure, provided you sew it up again along the tear when you are done. This distinction can be appreciated in terms of a knotted and an unknotted loop of string (Figure 7). These figures are themselves topologically equivalent, but it is quite impossible to deform one to look

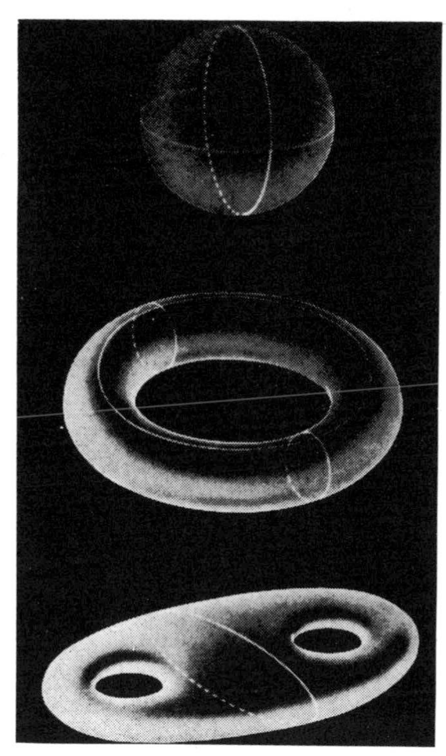

FIGURE 6. *The simplest closed surfaces in three-dimensional space: (top) the two-sphere; (middle) the torus; (bottom) the double torus. None of the closed curves shown on the torus or the double torus can be shrunk to a point without leaving the surface. On the other hand, every closed curve in the two-sphere can be shrunk to a point without leaving the surface. [Courtesy Addison-Wesley, Reading, Mass.]*

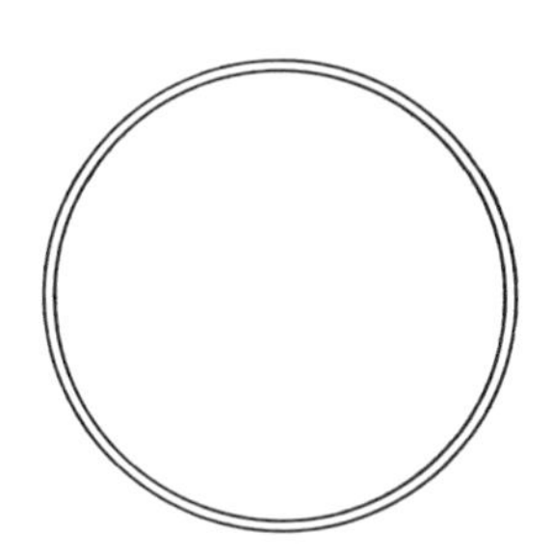

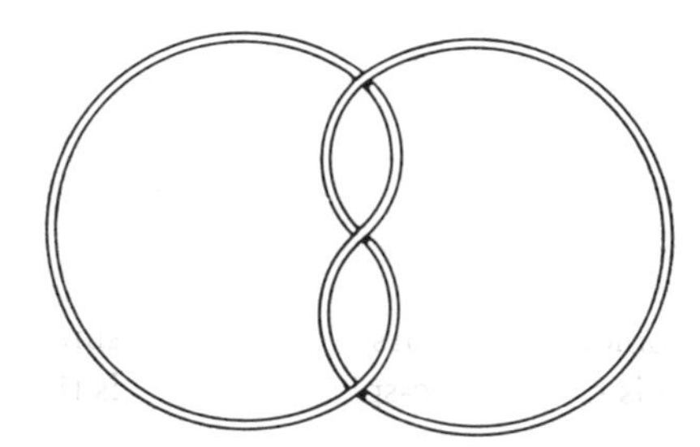

FIGURE 7. *A knotted loop and an unknotted loop are topologically equivalent, but neither can be deformed into the other unless a temporary cut is made.*

like the other. To deform the knotted loop into a circle you must temporarily cut it, untie the knot, and rejoin the ends.

Going back to surfaces, we know then that, if we encounter a closed surface in solving a differential equation in three-dimensional space, it must be topologically one of these. But more than that, it must be a surface which can be covered by a family of more or less parallel curves. A sphere cannot be covered with such a family of curves. Think, for example, of the equator and the circles of latitude on a globe. These cover the sphere neatly except for the poles, where the curves degenerate. It is easy to convince yourself by trials, and it can be rigorously proved, that there is no way to cover a sphere by curves without there being at least one point where the curves break down. In fact, of all the surfaces in our infinite list, only the torus can be covered by curves. Consequently, only the torus can appear as a closed integral surface for a three-dimensional differential equation.

These facts show that topology can provide information of definite value even in applied mathematics. Let us consider some of the questions which arise naturally when we review the facts about surfaces.

The simple sequence of surfaces described in the preceding paragraphs contains every possible kind of closed surface in three-dimensional space. What about nonclosed surfaces — that is, surfaces of infinite extent? These are also known, but the classification is quite complicated. What happens if we go to higher dimensions? If we look for two-dimensional closed surfaces in four-dimensional space, we find a new infinite list of possibilities. One of these is the famous Klein bottle (Figure 8), which is

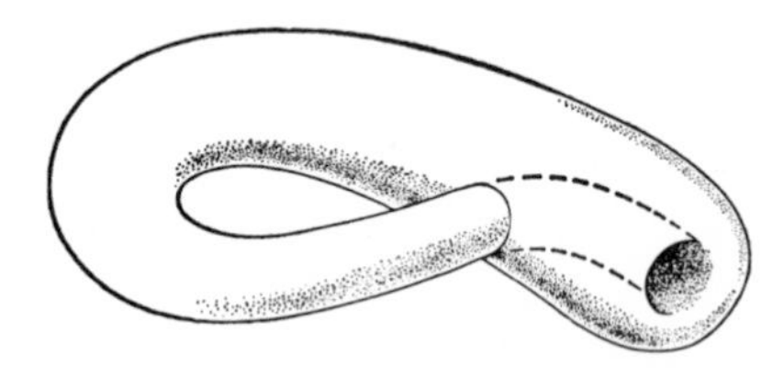

FIGURE 8. *The Klein bottle is a closed surface which does not fit into ordinary space without self-intersections.*

improperly displayed in three-space. It can be put in three-space only if you let it intersect itself. Of all the new surfaces, only the Klein bottle can be covered with curves, so it is the only new one that might appear in solving a four-dimensional differential equation. What happens in five-space? Nothing new. There are no further kinds of closed two-dimensional surfaces in any space.

Suppose we look for three-dimensional surfaces in higher-dimensional space. Can we again find a simple description of all the types? No, or at least not yet. There appears to be an overwhelming number of possibilities.

A truly remarkable problem appears here. The simplest type of closed three-dimensional surface is the three-sphere. This is the surface of a solid four-dimensional ball. It shares with its relative, the two-sphere, the property of being *simply connected.* In everyday terms this means that, if you have a loop of string in the surface of a ball, you can always gather it in to a point without removing it from the surface. Of all closed two-dimensional surfaces, only the two-sphere has this property (see Figure 6). Among closed three-dimensional surfaces the only known simply connected one is the three-sphere. Poincaré [3] conjectured about sixty years ago that no other closed three-dimensional surface has this property, but this has never been proved. A similar problem can be raised concerning spheres of higher dimension. Generally speaking, questions of this sort get harder as dimensions increase. It was a great surprise, therefore, when Smale [4], Stallings [5], and Zeeman [6] proved the generalized Poincaré conjecture for dimensions five and up. Eight years later the question is still open for both three and four dimensions.

Among two-dimensional surfaces, only the torus and Klein bottle can be smoothly covered by curves. What higher-dimensional surfaces can be so covered? Even though we don't know all the surfaces, we can decide this question for any explicitly given surface. On the other hand, we have no general method for deciding whether or not an explicit surface of high dimension can be paved with more or less parallel two-dimensional surfaces.

Differential Topology

The latest development in this area is the rise of a new subject called differential topology. It is hard to date its origin, but differential topology can be said to have begun with the work of Thom in the early 1950's [7].

Topologists do not distinguish between a square and a circle, because the one can be deformed into the other by stretching. The curves and surfaces which arise in analysis are smooth, like the circle, the sphere, or the torus; they have no corners or edges. Differential topology focuses entirely on these smooth curves and surfaces.

The differential topologist uses a more restrictive notion of equivalence of surfaces than the ordinary topologist. When a differential topologist deforms a smooth surface, he not only keeps it smooth, he keeps every smooth curve lying in that surface smooth. It is easier to describe what he does not do than what he does. Consider an ordinary plane and move it into itself as follows (Figure 9): Every point above a certain line L is kept fixed, but every point below L is moved to the right by an amount equal to its distance below L. Then any smooth curve which crosses L develops a corner after the plane is moved. This is the sort of thing that is prohibited by the differential topologist, although it would be accepted by the ordinary topologist. A differential topologist regards two smooth surfaces as equivalent if the first can be deformed in this more restrictive fashion to look like the second, possibly with some temporary tearing and

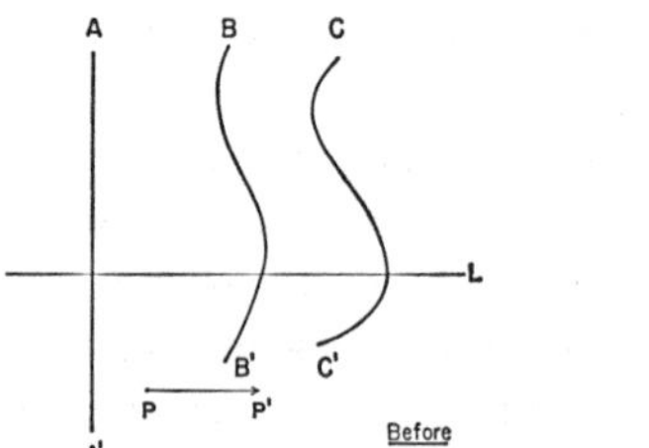

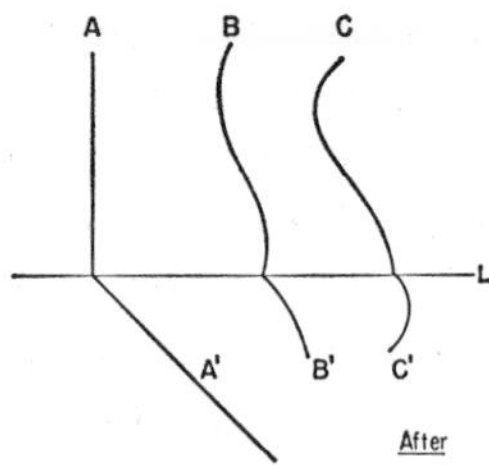

FIGURE 9. *A motion of the plane into itself which is not admissable in differential topology. Points on and above line* L *are fixed. Points below* L *are displaced to the right by an amount equal to their distance below* L*; for example,* P *is moved to* P′. *The effect of this transformation is to introduce corners on the previously smooth curves* AA′, BB′, *and* CC′, *at the points where they cross* L.

resewing. The rigorous formulation of these ideas involves the ideas of the differential calculus, and this is what gives the subject its name.

In low dimensions it is true that two smooth surfaces which are topologically equivalent are also equivalent in the sense of differential topology. But in 1955 Milnor gave an example of two seven-dimensional surfaces which are topologically equivalent but not differentiably equivalent [8]. This shows that differential topology is not just topology masquerading in new clothing, it is a genuinely new subject. In another sense, therefore, differential topology began with Milnor's example.

One of the greatest surprises in this now very popular field was found by Kervaire in 1960. He gave an example of a closed ten-dimensional surface with corners which is not topologically equivalent to any smooth surface [9].

Let us describe this surface by analogy in the geometric terms of our intuition. Since the figure in question is a surface, it can be smoothed out at any point. But no matter how we smooth it, it always has at least one corner. If we smooth out that corner, another one pops up automatically somewhere else.

Conclusion

In conclusion, let me point out that I have discussed this general field of mathematics entirely in terms of specific facts and problems. Unfortunately the methods involved in proving these facts involve such a long journey up the ladder of abstraction that it is impossible to give, in any brief article, a fair idea of how they work. Separated from the methods which establish them, facts can convey only a partial picture of mathematics. Like the great temples of some religions, mathematics may be viewed only from the outside by those uninitiated into its mysteries. Anyone who thinks at all about what is involved in asserting that Kervaire's ten-dimensional surface is unsmoothable will sense the power of the methods topologists have developed for organizing our knowledge of space, but understanding these methods is reserved for those who devote years to the study of mathematics.

BIBLIOGRAPHY

1. H. Poincaré, *Les méthodes nouvelles de la mécanique céleste*, Gauthier-Villars, Paris, 1892.
2. V. I. Arnol'd, *Dokl. Akad. Nauk SSSR, 145*, 487, 1962 (in Russian).
3. H. Poincaré, *Rend. Circ. Mat. Palermo 18*, 45, 1904.
4. S. Smale, *Bull. Am. Math. Soc.*, *66*, 373, 1960.
5. J. R. Stallings, *ibid.*, *66*, 485, 1960.
6. E. C. Zeeman, *ibid.*, *67*, 270, 1961.
7. R. Thom, *Compt. Rend.*, *237*, 1733, 1953.
8. J. Milnor, *Ann. Math.*, *64*, 399, 1956.
9. M. A. Kervaire, *Commentarii Math. Helv.*, *34*, 257, 1960.

BIOGRAPHICAL NOTE

Born in 1921, Andrew Gleason obtained his bachelor's degree in mathematics from Yale University in 1942. After four years of service in the Navy during World War II, he was elected to the Society of Fellows at Harvard University, and in 1950 he was appointed assistant professor of mathematics. Before he taught a single class, he was recalled to active duty in the Navy and served two more years. At that time he was primarily interested in topological groups and contributed a crucial part of the solution of Hilbert's celebrated "fifth problem." He is now a professor at Harvard, and his most recent work has been in the area of functional analysis. Professor Gleason has also been very active in the reform of school mathematics.

The world is a Tower of Babel, with three billion people speaking hundreds of languages that are unintelligible to most of their fellow men. Nevertheless, all languages have some common structural properties that are as basic as the orbits of electrons within atoms or the fact that human beings have precisely two eyes and two ears. Mathematical linguistics, a new science now flourishing in the United States and the U.S.S.R., attempts to analyze these basic properties. It promises to have direct applications to semantics and computer programming. Ultimately, it may lead to information processing in scientific articles and to language translation by computer.

Mathematical Linguistics

Zellig Harris

Over the last twenty years, mathematical linguistics, one of the youngest sciences, has been developed primarily in the United States and in Russia. It is an outgrowth of structural linguistics devoted to the application of mathematics to studying the structure of languages.

There are several hundred languages spoken by man, and there are scores of distinct language families. For instance, the several groups of languages spoken by American Indians differ from one another as greatly as the European languages differ from Chinese. There are languages spoken by hundreds of millions of people, for example Chinese and English, and there are distinct languages used by tribes comprising only a few villages. There are languages that have a literature developed over many centuries and are used for complicated theoretical and scientific discussions, and there are languages which have no written form. Within the existing languages, there are distinct sublanguages, for instance the technical jargon of science or the underworld slang in English, or the language used exclusively by women in certain primitive tribes. The study of this incredible variety of languages, their phonetics and grammar, and their change, is the task of descriptive and comparative linguistics. By the very nature of things, this study must be undertaken separately for each language or group of languages. It turns out, however — and this is one of the most important results of mathematical linguistics — that it is possible to isolate certain structural properties that are common to all languages. These properties are of a formal nature and can be described mathematically. Thus, we may talk about a basic mathematical structure of human language rather than of a particular language. In studying this structure, mathematical linguistics proceeds like every other discipline that uses mathematics to study a natural or social phenomenon. Certain observations are formulated in mathematical terms and taken as basic assumptions. Deductive reasoning is used to derive certain consequences of the assumptions, and the consequences are tested experimentally by comparing the theoretical predictions with the observed phenomena. In carrying out these observations, computers can play an important part. Yet it is important to distinguish mathematical linguistics from computational linguistics, which covers any processing of language data by use of computers. Computers are useful for mathematical linguistics but are not, in principle, indispensable. Here we see again the similarity between mathematical linguistics and other sciences that use mathematics.

This similarity extends also to the area of applications. Mathematical linguistics is a basic science, in which the primary motivation is the quest for knowledge: the desire to understand the nature of human language and, through this perhaps, the nature of thought. But this purely theoretical quest has important practical applications, for instance to mechanical translation and mechanical abstracting.

So far, the mathematical concepts used by linguistics have come mostly from algebra: mappings between sets, relation between mappings, partially ordered sets, semigroups. There are also certain graphs that appear in mathematical linguistics, and there are interesting connections with the theory of automata and with combinatorial mathematics. Stochastic processes appear, but in limited ways. The structure of formal languages and of computer programming languages is closely related to part of natural language structure. Mathematical logic serves as a tool in investigating and formulating linguistic material, rather than as an equivalent system (because of the absence in natural language of an equivalent to truth tables).

However, the debt of mathematical linguistics to mathematics is chiefly the attitude and the way of thinking rather than any particular result. It is perhaps no accident that most graduate students in this field come from mathematics or logic rather than from linguistics.

A mathematical description of language is possible because every language is a collection of discourses (utterances or writings) and every discourse is a sequence of discrete elements. It is a sequence of sentences, each of which is a sequence of words (or morphemes, i.e., stems and affixes), each of which is a sequence of sounds or letters. These elements are arbitrary; one could replace them by other elements without changing the linguistic structure. The particular sounds or letter shapes are important for recognition but are not related to the meanings of the words or their syntax (position in sentences); words which sound different may have the same meaning and syntax in different languages. And, as to the meanings, it has not proved possible to base a grammar on the meanings of words. Rather, grammars can be stated on the basis of how words can be related in respect to syntactic property. The great importance of stating which combinations form sentences and which do not is due to the necessary fact that in each language only certain sound sequences and word sequences constitute sentences; the others do not. Identifying those that do is the activity of grammar, which has to be investigated separately for each language. In structural and mathematical linguistics, however, we seek general methods for this investigation and general ways of characterizing the sequences that constitute sentences and the relations among these sequences.

An example of this is a general method for discovering the boundaries of words or morphemes without relying on their meanings. Suppose we are given a sufficiently large set of texts typed in a language unknown to us, without capital letters, without punctuation marks, and without any spaces between individual words. It is clearly a problem in cryptanalysis to deduce anything about the meaning of each text. It is empirically verified, however, that there is a method for dividing each text into words, and each word into morphemes (roots and affixes). There is also a method for classifying words syntactically and for finding the sentence boundaries among the words of a text. The methods are purely formal. The first method has been verified by having a computer perform the work.

The procedure in this method is rather simple. Thus, in a given sentence or text, we can find where the boundaries of words are located among the n successive letters or sounds (phonemes) of the sentence by comparing the given sentence with many other sentences or texts having the same first k letters, for each k from 1 to n. This is done as follows: For each k, we ask how many different letters occupy the $(k + 1)$th place in all the sentences that have the same first k letters. The points at which the number of different successors rises to a peak correspond to boundaries of morphemes and of words in the given sentence. This has been shown empirically by testing a great number of sentences. A computer test, however, is possible only for the more difficult case of morpheme boundaries within single words since we can put all English words in a computer for computing, but we cannot put millions of sentences into the computer memory. For example, the computer counts for the word *antithetically* are

$$\mathrm{a}_{26}\ \mathrm{n}_{22}\ \mathrm{t}_{13}\ \mathrm{i}_{23}\ \mathrm{t}_{9}\ \mathrm{h}_{4}\ \mathrm{e}_{7}\ \mathrm{t}_{1}\ \mathrm{i}_{1}\ \mathrm{c}_{26}\ \mathrm{a}_{1}\ \mathrm{l}_{26}\ \mathrm{l}_{1}\ \mathrm{y}_{26},$$

where the number after a particular letter is the number of different letters in all words (and two-word sequences) of English which have the same beginning up to that particular letter. The peak after "anti" indicates a morpheme boundary, as does the peak after "ic," etc. (Word boundary has the value 26 except in some cases where the grammar restricts the variety of following words.) The small rise after "e" here is in part due to the different variants of the morpheme *thet* (e.g., "antithesis"), and the morpheme boundary before "ic," which is missed here, is caught by the same method used backwards from the end of the word. In this way, because the variety of interword transitions is necessarily greater than the variety of intersound transitions within a word, we can segment a sentence or text into its morphemes purely on the basis of the sequences of letters.

A deeper and much more significant result refers to transformation of sentences and to the fact that the rules governing these transformations are the same for all languages. We explain first what is meant by one sentence being (transformationally) equivalent to another. Consider, for instance, sentences of the form N_1 VN_2, consisting of a noun N_1 followed by a verb V followed by a second noun N_2. For instance, this basic form can be made into a sentence in the following different ways.

The man took a book.	The man took a rain.
The book took a man.	The rain took a man.
The man took a walk.	The man took a wife.

Not all of these sentences are equally acceptable. A sample of the users of the language will agree that "The man took a book" is an acceptable sentence and that "The man took a walk" or "The man took a wife" are acceptable as metaphors. "The man took a rain" is not acceptable.

Consider now a sentence of the form: *It is* N_1 *who* VN_2. If we use the same words as we did in the sentence form N_1 VN_2, we obtain sentences of the same grading of acceptability. For instance, "It is the man who took the book" is as acceptable as "The man took the book"; "It is the man who took a walk" is as acceptable as a metaphor as "The man took a walk." The sentences, "The man took a rain" and "It is the man who took a rain," are equally unacceptable.

Using this observation, one can define the equivalence of two sentence forms. The two forms are equivalent if the sentences resulting from replacing the variables by words in the two forms always have the same degree of acceptability.

Given two equivalent sentence forms, *a* and *b*, it is possible to distinguish which of them is more complicated, for instance in the sense that it involves more words. Suppose that the more complicated one is *b*; we say in this case that *b* is obtained from *a* by a transformation. There is a transformation, call it *t*, which transforms the sentence form N_1VN_2 into *It is* N_1 *who* VN_2. The transformation *t* transforms the acceptable sentence, "The man took a book," into "It is the man who took the book," and the unacceptable sentence, "The man took a rain," into the unacceptable sentence, "It is the man who took a rain." The whole system of sentences and transformations of sentences in languages can be described in mathematical terms, and it turns out that, in all the languages investigated thus far, all possible transformations (of which there are about one hundred major ones) fall into some five structural classes which are much the same in all languages. Also, in all languages thus far investigated, one finds the same kind of "kernel sentences," that is, sentences which are not transformations of simpler ones. It is possible to take any sentence in a language and determine a set of kernel sentences from which the given sentence could have been obtained by successive transformations. In general, this can be done in one way only. If the decomposition is possible in more than one way, that is, if the sentence could have been derived by transformation of two or more kernels, it turns out that the sentence is ambiguous.

Here is an example of such a decomposition (of an unambiguous sentence). The sentence reads:

> Adrenaline probably intervenes in the breakdown of proteins and lipids.

The elementary sentences from which this sentence is obtained are "protein breaks" and "lipid breaks." The transformations leading from

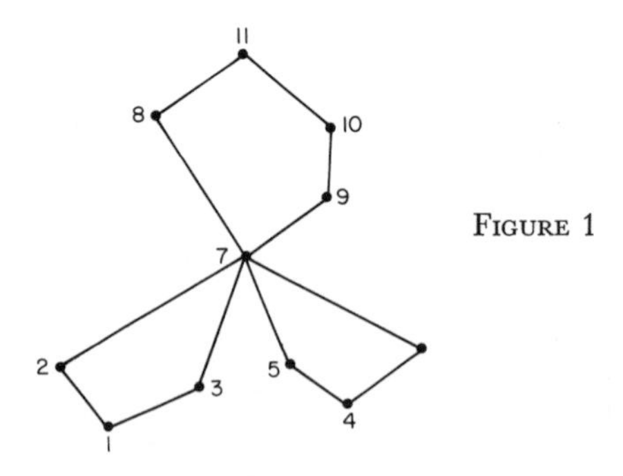

FIGURE 1

the elementary sentences to the final sentences are diagrammed as the set shown in Figure 1.

1 and 4: Elementary sentences "protein breaks" and "lipid breaks."
2 and 5: Plural, on first word (noun) of each operand (1 and 4).
3 and 6: Adverb "down," on second word of each operand.
7: "And" linking sentences "proteins break down" and "lipids break down."
8: Elimination of parallel material to yield "proteins and lipids break down."
9: "Adrenaline intervenes in, "operating on the product of transformation 7.

10: Adverb "probably" operating on "intervenes."
11: Period indicating that sentence is complete.

(Note that this set of transformations is only partially ordered. It is a flow chart indicating that some transformations must precede others but that in some cases the order is not critical. For example, 9 must precede 10, but 8 can be applied in parallel with 9 and 10.)

A reader looking at this may have the impression it has been done by sight by a person familiar with the meaning of the words. It is important to remember that this is not so. The decomposition can be performed mechanically on a computer. Furthermore, if the sentence is translated into another language, say Korean (a language with which we really experimented), and the resulting translated sentence is again decomposed mechanically into elementary sentences, we obtain the elementary sentences that are (roughly) the translations into Korean of the kernel sentences listed before.

We have seen that it is useful to view language as a set of sentences combined with a set of transformations of these sentences. This point of view can be explored much further. It is possible to formalize various properties of a language and its sublanguages (scientific language, the language of grammar, and so on) by studying subsets of sentences and the way transformations act on these sentences. This leads to mathematical structures interestingly different from those studied in mathematical logic and mathematics. This very abbreviated review shows that mathematical linguistics today involves basic mathematical approaches rather than application of detailed mathematical techniques. This may and probably will change in the future.

The results of structural and mathematical linguistics have various applications. They provide explications or solutions to many problems involving language, sentence, meaning, and the like, for example that ambiguity and paraphrase arise from particular types of transformation succession in a lattice or that one can determine the presumably intended meaning of an ambiguous sentence by relating its decompositions to those of neighboring sentences in the discourse. They provide necessary material for theories of discourse and of semantic information, for example that information involves the repeating of words, and discourse involves the repeating of kernel sentences. They define a normal form (i.e., a unique form reachable by formal operations) for each sentence of the language in such a way that the similarities and differences among sentences can be stated in a unique way that correlates with their informational relations. They provide methods for computer analysis of each sentence into kernel sentences plus transformations, or into word strings. This has been done in various ways, including a simple cycling automaton and a single scan that gives a picture of what can be understood serially at each successive word of the sentence. These linguistic methods of analyzing serial events (sequences), once they are mathematically formulated so as to be independent of the language data, may be of interest in other fields, for example, for analyzing computer operations or linear chemical formulas.

Because of the close correspondence between the normalization of a sentence and the information in the sentence, it may be possible to devise computer processing and comparison of the information contained in scientific articles, and it may be possible to see something of the special structure of the language of science.

BIBLIOGRAPHY

L. Bloomfield, *Language*, Holt, Rinehart & Winston, New York, 1933.

N. Chomsky, *Syntactic Structures*, Mouton, The Hague, 1957.

H. Hiz, *The Role of Paraphrase in Grammar*, TDAP series, No. 53, University of Pennsylvania, Department of Linguistics, Philadelphia, 1963.

BIOGRAPHICAL NOTE

Zellig Harris is one of America's foremost linguists. He is the author of several books, in particular *Methods in Structural Linguistics*, University of Chicago Press, Chicago, 1951, a book that played a decisive role in the history of structural linguistics, and *Mathematical Structures of Languages*, *Interscience Tracts in Mathematics*, Wiley, New York, 1968. Besides his linguistic work, which has always been strongly influenced by his early studies of mathematics, Harris has also had professional interests in social research. He is a professor of linguistics at the University of Pennsylvania and often a visiting professor of an Israeli kibbutz.

Most classical mathematics, from elementary algebra through differential equations, deals with the world as if all objects and all events were continuous. In many situations common in physics, chemistry, and other sciences, however, the only realistic explanation is in terms of collections of things acting one step at a time in combinations. The mathematics appropriate to this kind of thinking is called *combinatorial analysis*. Although its concepts are generally easy to understand, work in this field is extremely difficult. Some of the most interesting problems in combinatorial analysis have been posed as clever puzzles that challenge mathematicians and nonmathematicians alike. Although some of them may seem almost frivolous at first glance, they almost always have important and direct applications to concrete scientific problems.

Combinatorial Analysis

Gian-Carlo Rota

Combinatorial analysis — or, as it is coming to be called, combinatorial theory — is both the oldest and one of the least developed branches of mathematics. The reason for this apparent paradox will become clear toward the end of the present account.

The vast and ill-defined field of applied mathematics is rapidly coming to be divided into two clear-cut branches with little overlap. The first covers the varied offspring of what in the past century was called "analytical mechanics" or "rational mechanics," and includes such time-honored and distinguished endeavors as the mechanics of continua, the theory of elasticity, and geometric optics, as well as some modern offshoots such as plasmas, supersonic flow, and so on. This field is rapidly being transformed by the use of high-speed computers.

The second branch centers on what may be called "discrete phenomena" in both natural science and mathematics. The word "combinatorial," first used by the German philosopher and scientist G. W. Leibniz in a classic treatise, has been in general use since the seventeenth century. Combinatorial problems are found nowadays in increasing numbers in every branch of science, even in those where mathematics is rarely used. It is now becoming clear that, once the life sciences develop to the stage at which a mathematical apparatus becomes indispensable, their main support will come from combinatorial theory. This is already apparent in those branches of biology where the wealth of experimental data is gradually allowing the construction of successful theories, such as molecular biology and genetics. Physics itself, which has been the source of so much mathematical research, is now faced, in statistical mechanics and such fields as elementary particles, with difficult problems that will not be surmounted until entirely new theories of a combinatorial nature are developed to understand the discontinuous structure of the molecular and subatomic worlds.

To these stimuli we must again add the impact of high-speed computing. Here combinatorial theories are needed as an essential guide to the actual practice of computing. Furthermore, much interest in combinatorial problems has been stimulated by the possibility of testing on computers heretofore inaccessible hypotheses.

These symptoms alone should be sufficient to forecast an intensification of work in combinatorial theory. Another indication, perhaps a more important one, is the impulse from within mathematics toward the investigation of things combinatorial.

The earliest glimmers of mathematical understanding in civilized man were combinatorial. The most backward civilization, whenever it let fantasy roam as far as the world of numbers and geometric figures, would promptly come up with binomial coefficients, magic squares, or some rudimentary classification of solid polyhedra. Why then, given such ancient history, is combinatorial theory just now beginning to stir itself into a self-sustaining science? The reasons lie, we believe, in two very unusual circumstances.

The first is that combinatorial theory has been the mother of several of the more active branches of today's mathematics, which have become independent sometimes at the cost of a drastic narrowing of the range of problems to which they can be applied. The typical — and perhaps the most successful — case of this is algebraic topology (formerly known as combinatorial topology), which, from a status of little more than recreational mathematics in the nineteenth century, was raised to an independent geometric discipline by the French mathematician Henri Poincaré, who displayed the amazing possibilities of topological reasoning in a series of memoirs written in the latter part of his life. Poincaré's message was taken up by several mathematicians, among whom were outstanding Americans such as Alexander, Lefschetz, Veblen, and Whitney. Homology theory, the central part of contemporary topology, stands today, together with quantum mechanics and relativity theory, as one of the great achievements in pure thought in this century, and the first that bears a peculiarly American imprint. The combinatorial problems that topology originally set out to solve are still largely unsolved (for example, the four-color problem to be discussed soon or Poincaré's conjecture, discussed in A.

Gleason's essay on differential topology). Nevertheless, algebraic topology has been unexpectedly successful in solving an impressive array of long-standing problems ranging over all mathematics. And its applications to physics have great promise.

What we have written of topology could be repeated about a number of other areas in mathematics. This brings us to the second reason that combinatorial theory has been aloof from the rest of mathematics (and that sometimes has pushed it closer to physics or theoretical chemistry). This is the extraordinary wealth of unsolved combinatorial problems, often of the utmost importance in applied science, going hand-in-hand with the extreme difficulty found in creating standard methods or theories leading to their solution. Yet relatively few men choose to work in combinatorial mathematics compared with the numbers active in any of the other branches of mathematics that have held the stage in recent years. One is reminded of a penetrating remark by the Spanish philosopher José Ortega y Gasset, who, in commenting upon the extraordinary achievements of physics, added that the adoption of advanced and accomplished techniques made possible "the use of idiots" in doing successful research work. While many scientists of today would probably shy away from such an extreme statement, it is nevertheless undeniable that research in one of the better developed branches of mathematics was often easier, especially for the beginner, than original work in a field like combinatorial theory, where sheer courage and a strong dose of talent of a very special kind are indispensable.

Thus, combinatorial theory has been slowed in its theoretical development by the very success of the few men who have solved some of the outstanding combinatorial problems of their day, for, just as the man of action feels little need to philosophize, so the successful problem-solver in mathematics feels little need for designing theories that would unify, and thereby enable the less talented worker to solve, problems of comparable and similar difficulty. But the sheer number and the rapidly increasing complexity of combinatorial problems have made this situation no longer tolerable. It is doubtful that one man alone could solve any of the major combinatorial problems of our day.

Challenging Problems

Fortunately, most combinatorial problems can be stated in everyday language. To give an idea of the present state of the field, we have selected a few of the many problems that are now being actively worked upon. Each of the problems has applications to physics, to theoretical chemistry, or to some of the more "businesslike" branches of discrete applied mathematics such as programming, scheduling, network theory, or mathematical economics.

1. *The Ising Problem*

A rectangular ($m \times n$)-grid is made up of unit squares, each colored either red or blue. How many different color patterns are there if the number of boundary edges between the red squares and the blue squares is prescribed?

This frivolous-sounding question happens to be equivalent to one of the problems most often worked upon in the field of statistical mechanics. The issue at stake is big: It is the explanation of the macroscopic behavior of matter on the basis of known facts at the molecular or atomic levels. The Ising problem, of which the above statement is one of many equivalent versions, is the simplest model that exhibits the macroscopic behavior expected from certain natural assumptions at the microscopic level.

A complete and rigorous solution of the problem was not achieved until the last five years, although the main ideas were initiated many years before. The three-dimensional analog of the Ising problem remains unsolved in spite of many attacks.

2. *Percolation Theory*

Consider an orchard of regularly arranged fruit trees. An infection is introduced on a few trees and spreads from one tree to an adjacent one with probability p. How many trees will be infected? Will the infection assume epidemic proportions and run through the whole orchard, leaving only isolated pockets of healthy trees? How far apart should the trees be spaced to ensure that p is so small that any outbreak is confined locally?

Consider a crystalline alloy of magnetic and nonmagnetic ions in proportions p to q. Adjacent magnetic ions interact, and so clusters of different sizes have different magnetic susceptibilities. If the magnetic ions are sufficiently concentrated, infinite clusters can form, and at a low enough temperature long-range ferromagnetic order can spread through the whole crystal. Below a certain density of magnetic ions, no such ordering can take place. What alloys of the two ions can serve as permanent magnets?

It takes a while to see that these two problems are instances of one and the same problem, which was brilliantly solved by Michael Fisher, a British physicist now at Cornell University. Fisher translated the problem into the language of the theory of graphs and developed a beautiful theory at the borderline between combinatorial theory and probability. This theory has now found application to a host of other problems. One of the main results of percolation theory is the existence of a critical probability p_c in every infinite graph G (satisfying certain conditions which we omit) that governs the formation of infinite clusters in G. If the probability p of spread of the "epidemic" from a vertex of G to one of its nearest neighbors is smaller than the critical probability p_c, no infinite clusters will form, whereas if $p \geqq p_c$, infinite clusters will form. Rules for computing the critical probability p_c were developed by Fisher from ingenious combinatorial arguments.

3. *The Number of Necklaces, and Polya's Problem*

Necklaces of n beads are to be made out of an infinite supply of beads in k different colors. How many distinctly different necklaces can be made?

This problem was solved quite a while ago, so much so that the priority is in dispute. Letting the number of different necklaces be $c(n, k)$, the formula is

$$c(n, k) = \frac{1}{n} \sum_{d|n} \phi(d) k^{n/d}.$$

Here, ϕ is a numerical function much used in number theory, first introduced by Euler. Again, the problem as stated sounds rather frivolous and seems to be far removed from application. And yet, this formula can be used to solve a difficult problem in the theory of Lie algebras, which in turn has a deep effect on contemporary physics.

The problem of counting necklaces displays the typical difficulty of enumeration problems, which include a sizable number of combinatorial problems. This difficulty can be described as follows. A finite or infinite set S of objects is given, and to each object an integer n is attached — in the case of necklaces, the number of beads — in such a way that there are at most a finite number a_n of elements of S attached to each n. Furthermore, an equivalence relation is given on the set S — in this case, two necklaces are to be considered equivalent, or "the same," if they differ only by a rotation around their centers. The problem is to determine the number of equivalence classes, knowing only the integers a_n and as few combinatorial data as possible about the set S.

This problem was solved by the Hungarian-born mathematician George Polya (now at Stanford) in a famous memoir published in 1936. Polya gave an explicit formula for the solution, which has since been applied to the most disparate problems of enumeration in mathematics, physics, and chemistry (where, for example, the formula gives the number of isomers of a given molecule).

Polya's formula went a long way toward solving a great many problems of enumeration, and is being applied almost daily to count more and more complicated sets of objects. It is nevertheless easy to give examples of important enumeration problems that have defied all efforts to this day, for instance the one described in the next paragraph.

4. *Nonself-intersecting Random Walk*

The problem is to give some formula for the number R_n of random walks of n steps that never cross the same vertex twice. A random walk on a flat rectangular grid consists of a sequence of steps one unit in length, taken at random either in the x- or the y-direction, with equal probability in each of the four directions. Very little is known about this problem, although physicists have amassed a sizable amount of numerical data. It is likely that this problem will be at least partly solved in the next few years, if interest in it stays alive.

5. *The Traveling Salesman Problem*

Following R. Gomory, who has done some of the deepest work on the subject, the problem can be described as follows. "A traveling salesman is interested in only one thing, money. He sets out to pass through a number of points, usually called cities, and then returns to his starting point. When he goes from the ith city to the jth city, he incurs a cost c_{ij}. His problem is to find that tour of all the points (cities) that minimizes the total cost."

This problem clearly illustrates the influence of computing in combinatorial theory. It is obvious that a solution exists, because there is only a finite number of possibilities. What is interesting, however, is to determine the minimum number $S(n)$ of steps, depending on the number n of cities, required to find the solution. (A "step" is defined as the most elementary operation a computer can perform.) If the number $S(n)$ grows too fast (for example if $S(n) = n!$) as the integer n increases, the problem can be considered unsolvable since no computer will be able to handle the solution for any but small values of n. By extremely ingenious arguments, it has been shown that $S(n) \leqq cn^2 2^n$, where c is a constant, but it has not yet been shown that this is the best one can do.

Attempts to solve the traveling salesman problem and related problems of discrete minimization have led to a revival and a great development of the theory of polyhedra in spaces of n dimensions, which lay practically untouched — except for isolated results — since Archimedes. Recent work has created a field of unsuspected beauty and power, which is far from being exhausted. Strangely, the combinatorial study of polyhedra turns out to have a close connection with topology, which is not yet understood. It is related also to the theory behind linear programming and similar methods widely used in business and economics.

The idea we have sketched, of considering a problem $S(n)$ depending on an integer n as unsolvable if $S(n)$ grows too fast, occurs in much the same way in an entirely different context, namely, number theory. Current work on Hilbert's tenth problem (solving Diophantine equations in integers) relies on the same principle and uses similar techniques.

6. *The Coloring Problem*

This is one of the oldest combinatorial problems and one of the most difficult. It is significant because of all the work that has been done to solve it and the unexpected applications that this work has often had to other problems. The statement of the problem is deceptively simple: Can every planar map (every region is bounded by a polygon with straight sides) be colored with at most four colors, so that no two adjacent regions are assigned the same color?

Most of the early attempts at solving this problem (up to around 1930) were based on direct attack, and they not only failed but did not even contribute any useful mathematics. Thanks to the initiative of H. Whitney of the Institute for Advanced Study and largely to the work of W. T. Tutte (English-Canadian) a new and highly indirect approach to the coloring problem is being developed, called "combinatorial geometry" (or sometimes the theory of matroids). This is the first theory of a general character that has been completely successful in understanding a variety of combinatorial problems. The theory is based on a generalization of Kirchhoff's laws of circuit theory in a completely unforeseen — and untopological — direction. The basic notion is a closure relation with the MacLane-Steinitz exchange property. The exchange property is a relation, $A \rightarrow \overline{A}$, defined on all subsets A of a set S such that, if x and y are elements of S and $x \in \overline{A \cup y}$ but $x \notin \overline{A}$, then $y \in \overline{A \cup x}$. In general, one does not have $\overline{A \cup B} = \overline{A} \cup \overline{B}$, so that the resulting structure, called a combinatorial geometry, is not a topological space. The theory bears curious analogies with both point-set topology and linear algebra and lies a little deeper than either of them.

The most striking advance in the coloring problem is a theorem due to Whitney. To state it, we require the notion of "planar graph," which is a collection of points in the plane, called vertices, and nonoverlapping straight-line segments, called edges, each of them joining a pair of vertices. Every planar graph is the boundary of a *map* dividing the plane into *regions*. Whitney makes the following assumptions about the planar graph and the associated map: (a) Exactly three boundary edges meet at each vertex; (b) no pair of regions, taken together with any boundary edges separating them, forms a multiply connected region; (c) no three regions, taken together with any boundary edges separating them, form a multiply connected region. Under these assumptions, Whitney concludes that it is possible to draw a closed curve that passes through each region of the map once and only once. An example is given in Figure 1. Whitney's theorem has found many applications since it was discovered.

7. *The Pigeonhole Principle and Ramsey's Theorem*

We cannot conclude this brief list of combinatorial problems without giving a typical example of combinatorial argument. We have chosen a little-known theorem of great beauty, whose short proof we shall give in its entirety. The lay reader who can follow the proof on first reading will have good reason to consider himself combinatorially inclined.

THEOREM: Given a sequence of $(n^2 + 1)$ distinct integers, it is possible

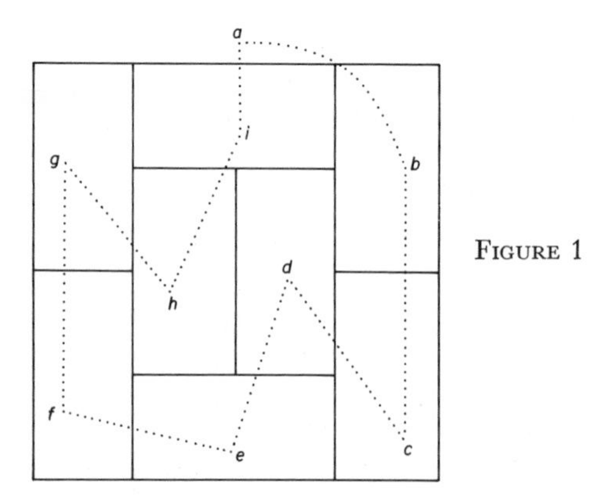

FIGURE 1

to find a subsequence of $(n + 1)$ entries which is either increasing or decreasing.

Before embarking upon the proof, let us see some examples. For $n = 1$, we have $n^2 + 1 = 2$ and $n + 1 = 2$; the conclusion is trivial since a sequence of two integers is always either increasing or decreasing. Let $n = 2$, so that $n^2 + 1 = 5$ and $n + 1 = 3$, and say the integers are 1, 2, 3, 4, 5. The theorem states that no matter how these integers are arranged, it is possible to pick out a string of at least three (not necessarily consecutive) integers that are either increasing or decreasing, for example,

$$1\ 2\ 3\ 4\ 5.$$

The subsequence 1 2 3 will do (it is increasing). Actually, in this case every subsequence of three elements is increasing. Another example is

$$3\ 5\ 4\ 2\ 1.$$

Here all increasing subsequences, such as 3 4 and 3 5, have at most two integers. There is, however, a wealth of decreasing subsequences of three (or more) integers such as 5 4 2, 5 2 1.

One last example is

$$5\ 1\ 3\ 4\ 2.$$

Here there is one increasing subsequence with three integers, namely 1 3 4, and there are two decreasing subsequences with three integers, namely 5 3 2 and 5 4 2; hence, the statement of the theorem is again confirmed.

Proceeding in this way, we could eventually verify the statement for all permutations of five integers. There are altogether $5! = 120$ possibilities. For $n = 3$, we have to take $n^2 + 1 = 10$ integers, and the amount of work to be done to verify the conjecture case by case is overwhelming since the possibilities total $10! = 3{,}628{,}800$. We begin to see that an argument of an altogether different kind is needed if we are to establish the conclusion for all positive integers n.

The proof goes as follows. Let the sequence of integers (in the given order) be

$$a_1, a_2, a_3, \ldots, a_{n^2+1}. \tag{1}$$

We are to find a subsequence of Sequence 1, which we shall label

$$a_{i_1}, a_{i_2}, \ldots, a_{i_{n+1}},$$

where the entries are taken in the same order as Sequence 1 but with one of the following two properties: either

$$a_{i_1} \leqq a_{i_2} \leqq \cdots \leqq a_{i_{n+1}} \tag{2}$$

or

$$a_{i_1} \geqq a_{i_2} \geqq \cdots \geqq a_{i_{n+1}}. \tag{3}$$

The argument is based on a *reductio ad absurdum*. Suppose that there is no subsequence of the type of Sequence 2, that is, no increasing subsequence of $(n + 1)$ or more entries. Our argument will then lead to the conclusion that, under this assumption, there must be a sequence of type of Sequence 3, that is, a decreasing sequence with $(n + 1)$ entries.

Choose an arbitrary entry a_i of Sequence 1, and consider all increasing subsequences of Sequence 1 whose first element is a_i. Among these, there will be one with a maximum number of entries. Say this number is l (= length). Under our additional hypothesis, the number l can be 1, 2, 3, . . . , or n, *but not* $n + 1$ or any larger integer.

We have, therefore, associated to each entry a_i of Sequence 4 an integer l between 1 and n; for example, $l = 1$ if all subsequences of two or more integers starting with a_i are decreasing. We come now to the crucial part of the argument. Let $F(l)$ be the number of entries of Sequence 1 with which we have associated the integer l, by the procedure just described. Then

$$F(1) + F(2) + F(3) + \cdots + F(n) = n^2 + 1. \tag{4}$$

Identity 4 is just another way of saying that with each one of the $(n^2 + 1)$ entries, a_i of Sequence 1 we have associated a number between 1 and l. We claim that *at least one* of the summands on the left-hand side of Identity 4 must be an integer greater than or equal to $n + 1$. For if this were not so, then we should have

$$F(1) \leqq n, \qquad F(2) \leqq n, \ldots, F(n) \leqq n.$$

Adding all these n inequalities, we should have

$$F(1) + F(2) + \cdots + F(n) \leqq \underbrace{n + n + \cdots + n}_{n \text{ times}} = n^2,$$

and this contradicts Identity 4, since $n^2 < n^2 + 1$. Therefore, one of the summands on the left-hand side of Identity 4 must be at least $n + 1$. Say this is the lth summand:

$$F(l) \geqq n + 1.$$

We now go back to Sequence 1 and see what this conclusion means. We have found $(n + 1)$ entries of Sequence 1, call them (in the given order)

$$a_{i_1}, a_{i_2}, \ldots, a_{i_{n+1}} \tag{5}$$

with the property that each one of these entries is the beginning entry of an increasing subsequence of l entries of Sequence 1 but is not the beginning entry of any longer subsequence of Sequence 1.

From this we can immediately conclude that Sequence 5 is a *decreasing* sequence. Let us prove, for example, that $a_{i_1} > a_{i_2}$. If this were not true, then we should have $a_{i_1} < a_{i_2}$. The entry a_{i_2} is the beginning entry of an increasing subsequence of Sequence 1 containing exactly l entries. It would follow that a_{i_1} would be the beginning entry of a sequence of $(l + 1)$ entries, namely a_{i_1} itself followed by the sequence of l entries starting with a_{i_2}. But this contradicts our choice of a_{i_1}. We conclude that $a_{i_1} > a_{i_2}$. In the same way, we can show that $a_{i_2} > a_{i_3}$, etc., and complete the proof that Sequence 5 is decreasing and, with it, the proof of the theorem.

Looking over the preceding proof, we see that the crucial step can be restated as follows: If a set of $(n^2 + 1)$ objects is partitioned into n or fewer blocks, at least one block shall contain $(n + 1)$ or more objects or, more generally, if a set of n objects is partitioned into k blocks and $n > k$, at least one block shall contain two or more objects. This statement, generally known as the "pigeonhole" principle, has rendered good service to mathematics. Although the statement of the pigeonhole principle is evident, nevertheless the applications of it are often startling. The reason for this is that the principle asserts that an object having a certain property exists, without giving us a means for finding such an object; however, the mere existence of such an object allows us to draw concrete conclusions, as in the theorem just proved.

Some time ago, the British mathematician and philosopher F. P. Ramsey obtained a deep generalization of the pigeonhole principle, which we shall now state in one of its forms. Let S be an infinite set, and let $P_l(S)$ be the family of all finite subsets of S containing l elements. Partition $P_l(S)$ into k blocks, say $B_1, B_2, \ldots, B_k$; in other words, every

l-element subset of S is assigned to one and only one of the blocks B_i for $1 \leq i \leq k$. Then there exists an infinite subset $R \subset S$ with the property that $P_l(R)$ is contained in one block, say $P_l(R) \subset B_i$ for some i, where $1 \leq i \leq k$; in other words, there exists an infinite subset R of S with the property that all subsets of R containing l elements are contained in one and the same of the B_i.

The Coming Explosion

It now seems that both physics and mathematics, as well as those life sciences that aspire to becoming mathematical, are conspiring to make further work in combinatorial theory a necessary condition for progress. For this and other reasons, some of which we have stated, the next few years will probably witness an explosion of combinatorial activity, and the mathematics of the discrete will come to occupy a position at least equal to that of the applied mathematics of continua, in university curricula as well as in the importance of research. Already in the last three years, the amount of research in combinatorial theory has grown to the point that a specialized journal is being published. In the past year, at least five textbooks and monographs in the subject have been published, and at least five more are now in print.

Before concluding this brief survey, we shall list the main subjects in which current work in combinatorial theory is being done. They are the following:

1. *Enumerative Analysis*, concerned largely with problems of efficient counting of (in general, infinite) sets of objects like chemical compounds, subatomic structures, simplicial complexes subject to various restrictions, finite algebraic structures, various probabilistic structures such as runs, queues, permutations with restricted position, and so on.
2. *Finite Geometries and Block Designs*. The work centers on the construction of finite projective planes and closely related structures, such as Hadamard matrices. The techniques used at present are largely borrowed from number theory. Thanks to modern computers, which allowed the testing of reasonable hypotheses, this subject has made great strides in recent years. It has significant applications to statistics and to coding theory.
3. *Applications to Logic*. The development of decision theory has forced logicians to make wide use of combinatorial methods.
4. *Statistical Mechanics*. This is one of the oldest and most active sources of combinatorial work. Some of the best work in combinatorial theory in the last twenty years has been done by physicists or applied mathematicians working in this field, for example in the Ising problem. Close connections with number theory, through the common medium of combinatorial theory, have been recently noticed, and it is very likely that the interaction of the two fields will produce striking results in the near future.

In conclusion, we should like to caution the reader who might gather the idea that combinatorial theory is limited to the study of finite sets. An infinite class of finite sets is no longer a finite set, and infinity has a way of getting into the most finite of considerations. Nowhere more than in combinatorial theory do we see the fallacy of Kronecker's well-known saying that "God created the integers; everything else is man-made." A more accurate description might be: "God created infinity, and man, unable to understand infinity, had to invent finite sets." In the ever-present interaction of finite and infinite lies the fascination of all things combinatorial.

BIOGRAPHICAL NOTE

Gian-Carlo Rota received his precollege education in Italy and in Ecuador, his bachelor's degree from Princeton University, and M.A. and Ph.D. degrees from Yale. Since then he has taught at various universities, principally at Massachusetts Institute of Technology, where he is now a professor of mathematics. In addition to combinatorial theory, his fields of interest include spectral theory and probability.

In a certain sense topology takes a downright primitive view of space, stripping geometry to its bare essentials. Some seemingly basic notions, such as straightness and size, often turn out to be irrelevant in topology, which is generally concerned with the way in which parts of geometric figures are connected. In its comparatively brief history, topology has been applied in many ways to problems in other branches of mathematics and, thus, to problems in the physical sciences. This essay illustrates the topological outlook and suggests some of the perplexities that have made it one of the most challenging fields of modern mathematics.

Point-Set Topology

R. H. Bing

Topology is a relatively new branch of mathematics. Those who took graduate training in mathematics thirty years ago did not have the opportunity to take a course in topology at many schools. Others had the opportunity but passed it by, thinking topology was one of those "new-fangled" things that was not here to stay. In that respect it was like the automobile.

One who is introduced to topology through popular lectures and entertaining expository articles may get the impression that topology is recreational mathematics. If he were to take a course in topology, expecting it to consist of cutting out pretty figures and stretching rubber sheets, he would be in for a rude awakening. If he pursued the subject further, however, he might be delighted to find that it is rich in substance and beauty. Of the references given at the end of this essay, Newman's book is the most pertinent to the material covered here. Some advanced texts on other subjects have a chapter treating the point-set topology used in their subjects. Some of these are very readable, for example Chapter 1 of Munroe.

In many respects, topology may be considered as an offshoot of geometry. From Euclid we have certain axioms for geometry. On the basis of some undefined terms, definitions, and axioms, certain theorems are proved. So it is with many treatments of topology. The undefined terms, definitions, and axioms that are used are different from those employed in geometry, but the methods are similar.

Topological Equivalence

In studies of the foundations of geometry, one may consider modifying certain of the axioms and seeing how this affects the resulting theorems. Around the end of the nineteenth century, Hilbert studied the effects of changing certain of Euclid's axioms. Veblen made further studies in this respect. One fundamental idea from plane geometry is the notion of betweenness. We use it to develop the notion of straightness. If this notion were modified in a certain way, the two objects shown in Figure 1 would

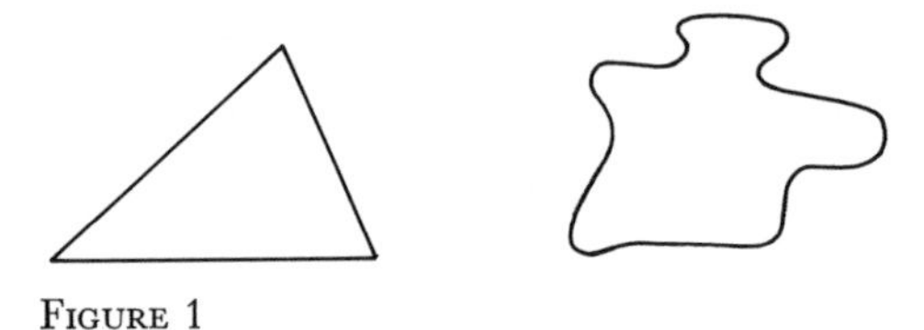

FIGURE 1

be alike. The first is a triangle and the second is a more general closed curve. In topology, they are considered alike, and we say they are *topologically equivalent*. The concept of topological equivalence corresponds in a way to congruence in Euclidean geometry.

Two sets, A and B, are called topologically equivalent if there is a one-to-one correspondence between them that is continuous both ways, as suggested in the following example. The curved arc A and the straight line segment B shown in Figure 2 are topologically equivalent. Although

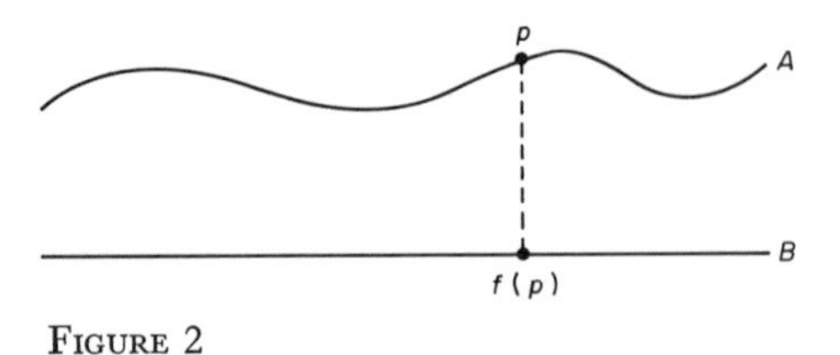

FIGURE 2

there are many one-to-one correspondences that could be chosen, the vertical projection of A onto B provides a convenient one; that is, a point p of A corresponds to the point $f(p)$ of B directly beneath it. This particular one-to-one correspondence from A to B is continuous because points close together in A go into points that are close together in B. The correspondence is also continuous the other way because points close together in B correspond to points close together in A. The expression "close together" is not precise, but this at least gives an intuitive explanation of continuity.

A one-to-one correspondence of a set A onto a set B that is continuous both ways is called a *homeomorphism* of A onto B. The projection of A onto B is an example of a homeomorphism of A onto B. Two sets are topologically equivalent if there is a homeomorphism of one onto the other. Accordingly, instead of saying that the two sets are topologically equivalent, we may say that they are homeomorphic.

It is possible to get a homeomorphism of a short segment I_1 onto a long segment I_2 (see Figure 3). The homeomorphism this time is given by a

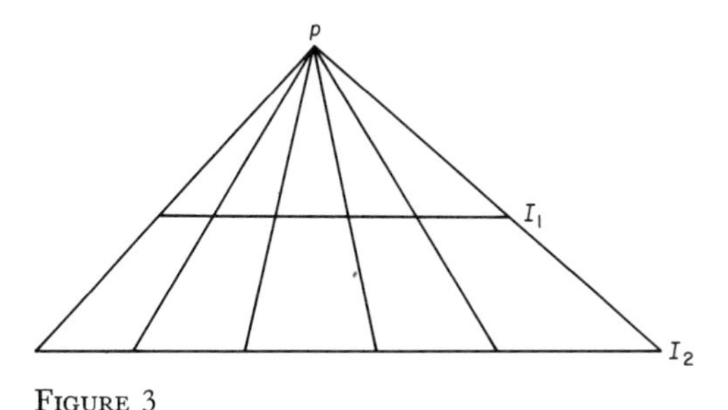

FIGURE 3

projection from a point. This homeomorphism shows that, in a certain sense, there are as many points on a short segment as on a long segment.

In Figure 4 we show some objects no two of which are homeomorphic. For example, the θ-curve is not topologically equivalent to the segment

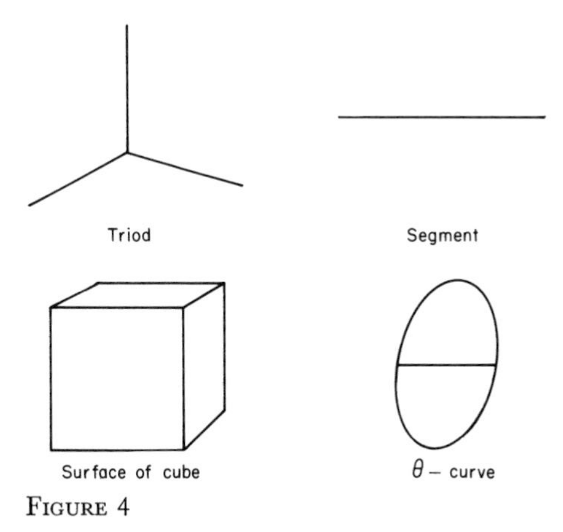

FIGURE 4

because no point of it could be made to correspond (in the proper way) to the end of the segment.

Pulling and Stretching Not Always Possible

Topology has been wrongly called "rubber sheet" geometry, owing to the misconception that, if two sets are topologically equivalent, it is always possible to deform one onto the other by "pulling and stretching but without breaking and tearing." If one figure can be deformed into another simply by pulling and stretching, the figures are topologically equivalent. But the converse of this statement is not true, as the following examples will show.

In E^3 (Euclidean 3-space, the familiar three-dimensional space of everyday experience), consider two sets each consisting of two tangent spheres; in the first set the spheres are tangent externally, and in the second set one sphere lies on the interior (except for the point of tangency) of the other. The two sets are topologically equivalent because there is a one-to-one correspondence of the proper sort between the two sets. However, it is not possible to deform one set onto the other in E^3 by "pulling and stretching but without breaking and tearing."

Another counterexample involves a cylinder or tube formed by sewing two opposite sides of a rectangular rubber sheet together. If the ends of the cylinder are sewed together, the resulting figure may resemble an inner tube. However, if a knot is tied in the cylinder before its ends are joined, the exterior view of the inner tube will be changed but its interior structure will not be. If after the knot is tied and before the final sewing

FIGURE 5

is done, one end of the cylinder is stretched and pulled over the knot before the ends are connected, another figure results. The three surfaces shown in Figure 5 are topologically equivalent but no one of them can simply be stretched to make it fit on any other one.

Complements of Figures

The "complement" of a geometric figure is the set of points not in the figure. For example, on the surface of the earth, the surface of the oceans, seas, lakes, and other bodies of water is the complement of the surface of the land. In the plane, the complement of a circle consists of two pieces: its interior and its exterior. The complement of a circle in E^3 would be all in one piece.

Even though two figures in the same space are topologically equivalent, their complements may be topologically different. In Figure 5, each of the surfaces has an interior and an exterior. It could be shown that the interior of the first is topologically equivalent to the interior of the second but not to the interior of the third. A small creature who knew about topological equivalence but not about straightness, length, and so forth, could not tell the difference between the inside of the first tube and the inside of the second. However, the exterior of the first tube in Figure 5 is like the exterior of the third but not like the exterior of the second. No two of the three figures have complements that are topologically equivalent.

J. W. Alexander described the "horned sphere" illustrated in Figure 6. It is topologically equivalent to the surface of a sphere, but its exterior is not like the exterior of a sphere.

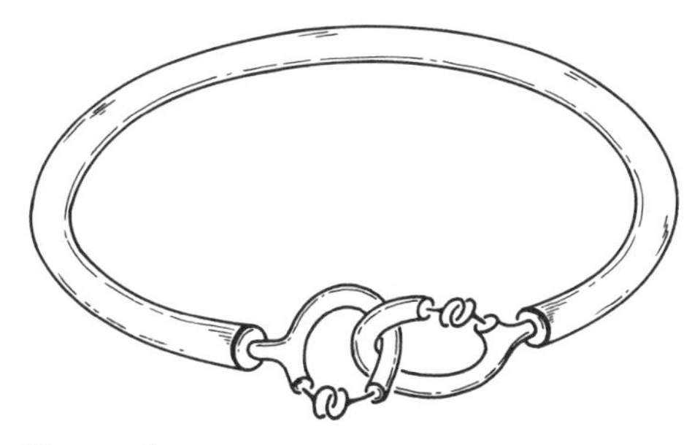

FIGURE 6

We might describe the horned sphere as follows. A long cylinder closed at both ends is folded until the ends are near each other and parallel. Then tubes are pushed out each end until they hook as shown. The process is continued by pushing out additional tubes, and so on. The resulting set has the property that, although it is topologically equivalent to the surface of a sphere, there is a circle in the exterior of the horned sphere that cannot be shrunk to a point without hitting the horned sphere. Thus, the complement of a horned sphere is not like the complement of a round sphere because each simple closed curve in the complement of a round sphere can be shrunk to a point without touching the sphere. (A "simple" curve in this context means a curve that does not cross itself.)

The interiors of circles and triangles appear to be alike. However, if one has a complicated simple closed curve such as shown in Figure 7, he might have difficulty in deciding if its interior is like the interior of a circle. If the curve were even more complicated, it might be impossible to tell merely by looking whether it even has an interior.

The Jordan curve theorem states that the complement of each simple closed curve J in the plane consists of two separate connected pieces and the curve is the boundary of each of them. One extension of the Jordan curve theorem due to Schoenflies may be stated as follows: If J_1 and J_2 are simple closed curves in the plane, there is a homeomorphism of the plane onto itself that takes J_1 onto J_2.

FIGURE 7

In plane analytical geometry, one studies translations and rotations of the plane onto itself — and perhaps even reflections through a line. These are examples of homeomorphisms of the plane onto itself. Under each of these particular homeomorphisms, any Figure F would go into a figure congruent to F. But the types of homeomorphisms used to show the truth of the foregoing extension of the Jordan curve theorem would frequently change the shapes of objects.

To those who think the theorem needs no proof since it is intuitively obvious, it may come as a surprise that the analogous theorem is not true in 3-space. Although the surface of a horned sphere described in previous paragraphs is topologically like the surface of a sphere (just as a simple closed curve is topologically like a circle), there is no homeomorphism of 3-space onto itself that will take the surface of a horned sphere onto the surface of a sphere.

The Jordan curve theorem is one of the more frequently used theorems of plane topology; it is used in such subjects as analysis or complex variables. It was not known for many years whether or not the Jordan curve theorem was true, though during most of this time many people suspected it to be true. Although several proofs of this theorem have now been given (and some only recently), these proofs are still complicated.

Some Unsolved Problems

There are many unsolved problems in topology, such as the famous four-color problem. Suppose a map is drawn on the surface of a sphere in such a way as to divide up this surface into regions, each with a boundary that is a simple closed curve. The four-color problem then asks if each such map on the surface of a sphere can be colored by use of no more than four colors. Regions whose boundaries meet only at a point are allowed to have the same color, but regions with a common boundary line are not allowed to have the same color.

Quite a few unsolved topology problems involve what are known as topological characterizations. Suppose we show first that a well-known space (such as a line, plane, or 3-space) satisfies a certain set of topological conditions and then that any space satisfying these same conditions must be topologically equivalent to the well-known space. We then say that these conditions constitute a *topological characterization* of the well-known space.

We have some good topological characterizations of the line. However, one of the interesting unsolved problems, called Souslin's problem, is in this area. It gives a certain set of conditions satisfied by the line and asks whether any space that satisfies them must be topologically equivalent to a line.

Several topological characterizations of the plane have been discovered during the past forty years, and a problem that was for many years unsolved asked whether or not a certain set of conditions constituted a topological characterization of the surface of a sphere.[1] The answer was finally given in the affirmative in 1946 (Kline sphere characterization).

Although topological characterizations of 3-space have been given, these are still complicated. Problems in this area are discussed by Bing in the reference given below. In view of the recent spurt toward increasing our knowledge of topological properties of 3-space, there is hope for improvements here.

BIBLIOGRAPHY

H. M. A. Newman, *Elements of the Topology of Plane Sets of Points*, Cambridge University Press, Cambridge, England, 1951.

M. E. Monroe, *Introduction to Measure and Integration*, Addison-Wesley, Cambridge, Mass., 1953.

R. H. Bing, "Some Aspects of the Topology of 3-Manifolds Related to the Poincaré Conjecture," *Lectures on Modern Mathematics*, Vol. II, T. L. Saaty, Ed., Wiley, New York, 1964.

BIOGRAPHICAL NOTE

R. H. Bing is a research professor at the University of Wisconsin and a leader of the active group of topologists working there. Wisconsin's topology program attracts many capable graduate students, and the lively topology seminars cause a flow of research topologists to spend their summers and research leaves at Wisconsin.

[1] For those who might be curious as to the precise form such a problem might take, even though they may not be able to understand all the technical terms, we give the complete statement: Is the connected compact metric space S topologically equivalent to the surface of a sphere if it has more than one point and satisfies the following conditions?

1. The neighborhoods of S are connected.
2. The complement of no simple closed curve is connected.
3. The complement of each pair of points is connected.

Dr. Bing's research interests have been chiefly in geometric topology. He did his graduate work at the University of Texas and, for a time, devoted much of his study to the topology of the plane, continuous curves, and abstract spaces. More recently, he has been interested in higher dimensional Euclidean spaces. He feels that the study of Euclidean spaces is applied in that it reveals properties of the real space in which we live.

A new field of science, called biomathematics, is beginning to emerge. Plenty of problems in modern biology would benefit from a mathematical approach, and mathematicians find the life sciences fascinating. This essay outlines some of the accomplishments in biomathematics, including research on the conduction of electric signals by nerves, which contributed to the winning of a Nobel Prize. Other topics include the flow of blood, calculations of radiation treatment for patients, and diffusion of radioactive tracers and other chemicals in the body. In some of this work, computers help; linked to instruments, they analyze data almost as fast as it is recorded. Who is best fitted to do biomathematics — a mathematician assigned to a biological or medical laboratory or a biologist with special training in mathematics? The answer has not yet been decided, but it is clear that men with an uncommon breadth of knowledge will be needed. So far, the mathematical techniques applied to biology have been conventional, but there is reason to suspect that ultimately biology will inspire the creation of new mathematics, as physics has so often done in the past.

Mathematics and the Biomedical Sciences

Hirsh Cohen

Several years ago, Eugene Wigner discussed "the unreasonable effectiveness of mathematics in the natural sciences." Mathematics has, indeed, been so successful in the physical sciences that it may be an understatement to call her either queen or handmaiden. As everyone knows, the affair has not developed so well with the biological sciences, where there has been almost an unreasonable ineffectiveness of mathematics. Wigner has observed that "the construction of machines, the functioning of which he can foresee, constitutes the most spectacular accomplishment of the physicist." The biologist is not seeking to construct; the machine is given to him, and he must analyze it. He would also like to foresee its future functioning, but he is constrained to systems as they exist. Perhaps this is why the understanding of biological phenomena in mathematical terms seems to have lagged.

This does not mean that mathematics has not entered at all into biological and medical problems. In virtually every area one finds some degree of quantitative formulation aided by various of the mathematical techniques. And this is not a recent occurrence; Euler wrote on both blood flow and mortality problems, and other important mathematicians before and after Euler — Descartes, Helmholtz, Volterra, Von Neumann — took up problems in physiology, ecology, and other kinds of biology. A glance through a series of medical handbooks reveals drug absorption formulas in pharmacology, growth-rate studies in morphology and embryology, not to mention nearly a century's worth of mathematical genetics. For many years, there has been work in special topics of mathematical biophysics. What is true, of course, is that mathematics has not played the central role in the biological sciences that it has in the physical sciences. Nor has biology acted to stimulate much new mathematical thought.

With the large-scale increase in quantitive measurement techniques and in experimental control, the situation should soon change. Data can now be acquired in flexible form at high speeds in many biological fields of research. For example, a small but effective real-time data-processing system in the laboratory of H. K. Hartline at the Rockefeller University not only enables dynamic light patterns presented to the eye of a horseshoe crab to be interpreted in terms of measured nerve impulse firings but permits very fast analyses of spectral and noise properties of the responses. With this kind of ability has come a great deal of data analysis and model formulation. This, in turn, has produced a number of mathematical theories and is sure to stimulate a good many more. The use of computers obviously has a great role in this movement. What has also appeared very clearly is the need to have more mathematicians engage in biological and medical problems and more biologists knowledgeable about mathematical formulations and techniques.

The briefly discussed examples that follow are intended to illustrate (not survey) biology and medicine as new fields of application for mathematics similar to that used in the physical sciences, not as applications of either statistical techniques or data-processing machinery. (Several texts and anthologies review computer applications, a topic that is not easily disentangled, especially since some degree of mathematical thinking inevitably underlies most computer usage.)

We have divided the mathematical work into three kinds of activities. In the first, we discuss some mathematical formulations that, having already stood serious testing, have qualified as important biological theories. In these, the mathematician's work is the further manipulation and extension of these formulations. The theory of the conduction of nerve impulses with its nonlinear diffusion equations, described a bit later on, is in this state. In the second group, we discuss problems that have not yet reached this stage. The mathematician's work is to assist in the formulation. He has the data, the biologist's theory, and his own formulative methods to work with, and he is engaged in model building. The model of iodine metabolism in the dog, to be discussed as an example of the problem formulation called compartment analysis, is such a case. Finally, in the third group, not discussed here, falls the mathematical work involved in the various kinds of data analysis and data handling. As one might expect, statistics, time series, and other pattern analyses are among the mathematical methods here.

An interesting observation is that those areas for which the mathematical treatment is most advanced, such as the nerve conduction problem or problems in arterial blood flow, are the ones closest to biological mechanisms and furthest from the applied biology of clinical medical practice. As we proceed from the application of mathematical techniques through mathematical formulation toward statistics and data analysis, we also thus move from pure biology toward medicine.

Will these mathematical forays into the biological sciences produce new mathematics? It remains to be seen. So far, although there are new particular problems, there are no new mathematical disciplines, and they do not yet seem to be required. One may imagine a large storehouse of mathematical tools ready for use and a much larger collection of extremely complex biological problems. What is now required is the formulation of these problems in mathematical terms, in itself an enormous task. As this is done, the need for newer mathematics may appear.

Brief Glimpses

Before we turn to a discussion of examples in moderate detail, we list very briefly a few of the many applications of mathematical science to biology and medicine that we shall *not* discuss:

1. Mathematical descriptions of "neural networks" have been suggestive, and sometimes enlightening, in the study of such large groups of interlinked nerve cells as the brain. The mathematical tools include the methods of logic and Boolean algebra.
2. For many years the flow of a viscous fluid in elastic tubes has been studied mathematically as a model of blood flow. Today this model is being improved to consider impedance matching at the branches of arteries and veins, and the flow of cells that barely pass through capillaries is being treated more accurately.
3. The use of potential theory in relating electric potentials on the surface of the heart to the electric voltages recorded on an electrocardiograph is now helping to clarify the effects of heart motion and chest wall thickness on such recordings.
4. Quantitative use of the ultracentrifuge requires the use of nonlinear diffusion equations of a nonstandard variety.
5. High-speed manipulation of complex Fourier series, today much faster than ever before, continues to be essential to x-ray crystallography, which has helped to elucidate the structure of such large molecules as hemoglobin, myoglobin, lysozyme, and chymotrypsinogen.
6. Quite useful computer routines can now calculate the distribution of radiation through the body of a patient who is being treated either with

several radiation beams or with implanted sources. These calculations allow us to tailor radiation treatment more closely to medical needs.

7. Understanding the growth cycle of white blood cells in both normal and malfunctioning blood systems is of great interest in leukemia. A variety of models is in use, and an attack on the problem of synchronous cell growth is now anticipated.

8. Models of the mechanical propulsion of single cells — by cilia, flagellae, or pseudopods — have been developed with reasonable success.

9. An interesting model, employing methods of algebraic topology (in particular, the theory of tolerance spaces), has been proposed to connect the physical structure and observable functioning of the brain. The theory is a qualitative one and so far is one of the few biomedical uses of modern geometry in its topological form.

Biological Theories and Mathematical Manipulation in Neurophysiology

Few biological problems have attained a mathematical description satisfactory enough so that manipulation becomes an interesting and productive study. Let us consider, first, several from a particular area, neurophysiology, in order to exhibit the kinds of problems that have arisen.

Conduction of Nerve Impulses

Information in the nervous system is carried in the form of electric pulses that travel along single nerve fibers. A set of equations that describe the conduction of these pulses was part of the Nobel Prize winning research by Hodgkin and Huxley.

Briefly stated, the voltage along the membrane of a single nerve fiber satisfies a nonlinear differential equation (see Table 1). With the exception of questions of multiple solution, to be discussed, this equation has been very successful in describing many of the aspects of nerve-fiber behavior. As a result, much work has been done to extend and generalize this model.

Certain nerve fibers of the squid, which are large enough to permit exceedingly precise and repeatable measurements, provided the first data for this model. In the squid, Equation 2 of Table 1 holds all along the length of the fiber; in vertebrates, the nerve fiber is covered for most of its length by a tissue called myelin, and the ionic exchange described by Equations 2 through 5 occurs only at nodal points. (For example, in certain nerve fibers of the frog, the distance between nodes is 2 millimeters (mm). The wavelength of the propagated pulse covers about 20 nodes.) In this case I_i is linearly proportional to $v(x, t)$ for the myelinated portion and follows Equation 2 for the nodes. The modified Hodgkin-Huxley equation has been tested for the myelinated fibers of frogs and again found to be very satisfactory.

More detailed numerical tests of this model show varying success:

1. If a step current stimulus is given and held constant, the amplitude of the step will determine pulse repetition rate. This is, in fact, the amplitude-to-frequency conversion that nerve cells use to convey information about intensities.

Table 1 Differential equations for nerve fiber behavior

The partial differential equation (a nonlinear cable equation):

$$C\left(\frac{\partial v}{\partial t}\right)_x = \frac{a}{2R}\left(\frac{\partial^2 v}{\partial x^2}\right)_t - I_i(v; m, n, h), \tag{1}$$

where a = radius of the fiber, C = effective capacitance of the membrane, and R = effective resistance of the cell material lying inside the membrane and where I_i, the ionic transmembrane current, arises from the flow of sodium, potassium, and other less important ions in and out of the nerve fiber (across the membrane).

The current and ionic flow equations: The current and the ionic flows satisfy

$$I_i = \bar{g}_{Na} m^3(v)h(v)(v - v_{Na}) + \bar{g}_K n^4(v)(v - v_K) + \bar{g}_l(v - v_l), \tag{2}$$

$$\frac{dm}{dt} = \frac{1}{\tau_m(v)}\,(m_\infty(v) - m), \tag{3}$$

$$\frac{dn}{dt} = \frac{1}{\tau_n(v)}\,(n_\infty(v) - n), \tag{4}$$

$$\frac{dh}{dt} = \frac{1}{\tau_h(v)}\,(h_\infty(v) - h). \tag{5}$$

The particular way in which m, n, and h appear in Equation 2 and the functions $\tau_m(v)$ and $m_\infty(v)$, and so on, have been obtained by direct experimental measurements on isolated nerve fibers.

2. The Hodgkin-Huxley system has been shown to have the property that, for current stimulus amplitudes just above threshold, a finite train of pulses appears. The number of pulses increases to infinity as the stimulus amplitude increases but then decreases again on further increase in stimulus amplitude, as is shown in Figure 1. This agrees basically with experimental results but differs in several important details:

a. the range of amplitudes for which the number of pulses is finite is very small, far smaller than what one could expect to obtain experimentally;

b. although experiment shows that intervals between pulses should decrease as amplitude of stimulus increases, this does not occur in any marked fashion in the calculated results.

Currently, attempts are under way to see whether the model gives the correct behavior of the nerve impulse as it branches off into nerve fibers that terminate in muscles. Whether the active propagation continues to

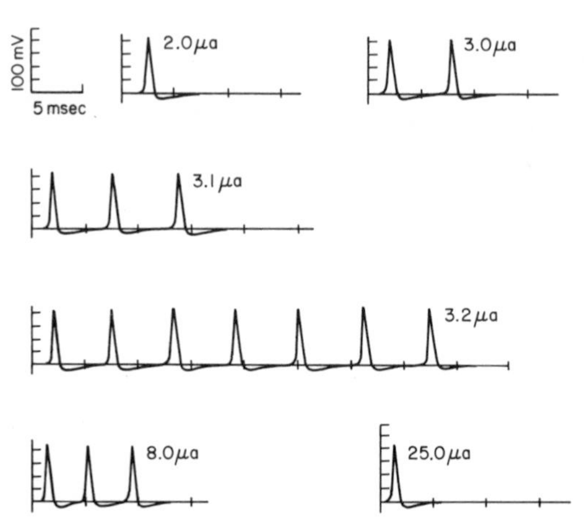

FIGURE 1. *Calculated response of the axon to a step current stimulus of the intensity indicated. The response is shown at a point two centimeters down the axon from the point of stimulus.*

the smaller fiber endings is now under experiment; appropriate calculations are also being carried out.

Multiplicity of Solutions

Both the mathematician and the physiologist are interested in how many solutions can be produced in nerves and how many satisfy the equations of Table 1. Since well-started nerve pulses are of fixed shape and travel at constant velocity along the nerve membrane, the appropriate solutions of the partial differential equations are certainly close to being self-sustaining traveling waves of the form $v(x, t) = v(x-\theta t)$. This assumption converts the partial differential equations into ordinary differential equations (which are to be solved subject to prescribed behavior very far from the pulse). These ordinary differential equations were originally solved by a choice of a likely value of θ, starting far toward one end of the range and integrating numerically toward the other end. Successive trials with different values of θ led to an extremely satisfactory pulse, one that compared well with experimental results in terms of its shape, its amplitude, and its velocity θ. However, another solution, not observed experimentally, was also found. (See Figure 2, in which the uppermost curve *a* is the usual nerve impulse.)

If the steady, traveling-wave assumption is not made, the initial-value problem itself for the original partial differential equations can be approximated by difference equations and, in this form, has been solved. In this way, the generation of the pulse can be observed, and also the manner in which the steady state develops can now be studied. It has been extremely satisfying to see that this analysis gives just the steady state found in

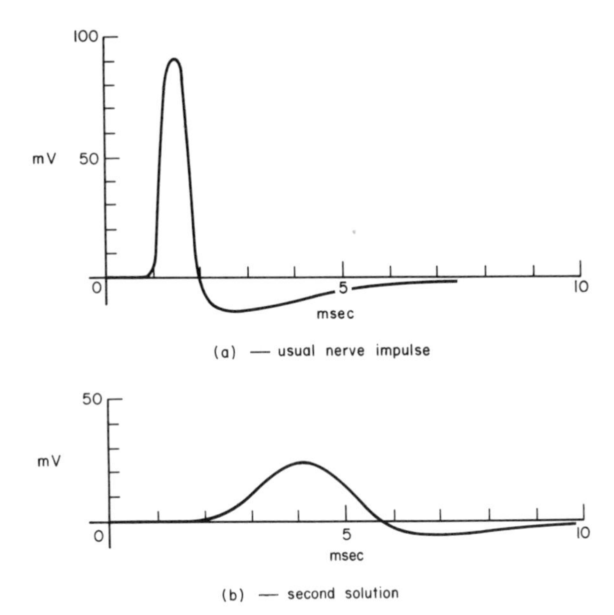

FIGURE 2. *Calculated pulse shapes.*

experiments (provided that the current stimulus is adequately above the threshold value) and predicted by the first analysis.

However, when the stimulus amplitude was very slightly below the threshold (and chemical conditions were normal), a pulse very much like curve *b* in Figure 2 was calculated. In this model, it did not continue very far but, instead, decayed. Thus, we must ask: Are *a* and *b* stable and unstable propagating solutions? Can we observe the unstable one on a nerve? We do not yet know the answer to the second question, though the reduction in pulse velocity by drugs such as procaine suggests that large enough doses of an appropriate drug, combined with a delicately adjusted stimulus, may slow down its decay until the less stable pulse can actually be observed.

In yet another use of the equations, the effects of drugs and poisons on *myelinated* fibers are being calculated. The idea is to make use of experimental, chemical, and electrical data with the model to compute drug effects on propagation velocities and other new characteristics.

At this point, and at a few others scattered through biology, the interaction with mathematics has become strong enough to begin to provide suggestions about how to search for new biological phenomena.

Sensory Reception

We have discussed how signals are conducted along nerve fibers, but we have not said how they are generated. The Hodgkin-Huxley model does not deal with this question. Let us turn to a famous series of experiments and a theory developed on data taken from the horseshoe crab, Limulus, in the laboratory of H. K. Hartline at the Rockefeller Institute. A useful review is given in a very recent book by Floyd Ratliff. The eye of Limulus is a compound eye, containing on the order of one thousand fixed receptors, the ommatidia (Figure 3). Each of these

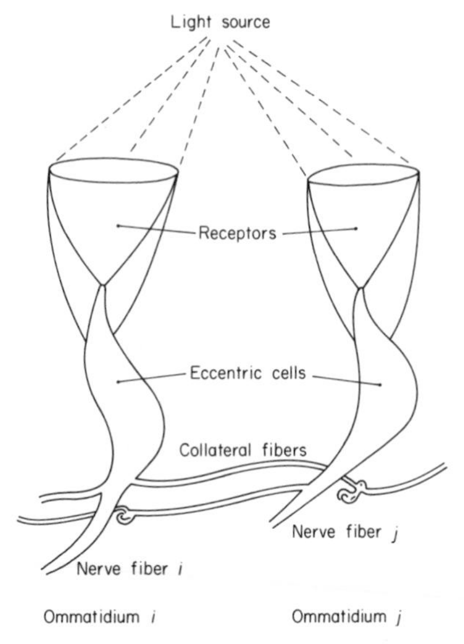

FIGURE 3

ommatidia is connected to a single nerve fiber. Each receptor-fiber pair represents a simple channel for the reception of light and the production of nerve impulses. One may think of the collection of ommatidia and other directly observable nerve fibers that link them together as representing a very primitive retina.

When a light beam falls on the lens of an ommatidium, it produces a volley of electric pulses in its nerve fiber. However, stimulating neighboring ommatidia makes a great difference in the rate at which these pulses occur. Such interdependence, called lateral inhibition, has been found in a wide variety of sensory systems.

In Limulus, stimulation of other receptors always weakens the signal from a given receptor. When the light pattern is kept constant, this decrease is expressed by the Hartline-Ratliff equation

$$r_i = e_i - \textstyle\sum_{i \neq j} K_{ij}(r_j - \tau_{ij}),$$

where it is understood that $(r_j - \tau_{ij})$ must be greater than zero. The τ_{ij} are thresholds that the actual rates (the r_j) in neighboring nerve fibers must exceed to have an effect. The K_{ij} are influence coefficients that measure the strength of this effect. Here r_i is the observed pulse rate in nerve fiber i, while e_i is the rate that would have been observed if only receptor i had been illuminated.

In experiments, the quantities that are measurable are r_i and e_i. If we mask all the neighboring ommatidia, light can be applied to a single ommatidium and e_i obtained. With the mask removed, r_i is obtained. This experiment has been done to find the interaction between pairs of ommatidia and has established quite firmly the presence of a threshold.

With only one other ommatidium uncovered, a particular term in the lateral inhibition equation can be studied. As the distance $i - j$ between the two ommatidia increases, the strength of interaction K_{ij} decreases and the threshold τ_{ij} increases (Figure 4). It appears, and this is still a con-

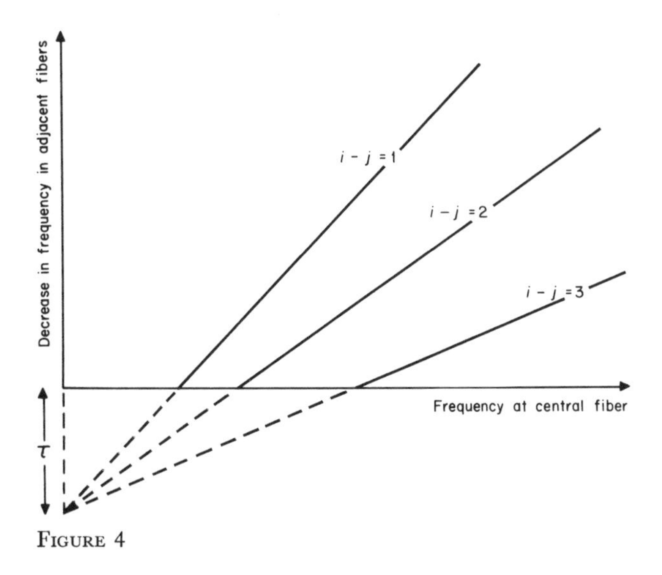

FIGURE 4

jecture, that the thresholds may in fact, be related to the K_{ij} simply: $\tau_{ij} = \tau/K_{ij}$, where τ is also as defined in Figure 4.

In an experiment where the light intensity is constant along each of a family of parallel lines, a model in which K_{ij} depends only on distance between ommatidia leads, when thresholds are neglected, to a linear integral equation:

$$r(x) = e(x) - \int_{-\infty}^{+\infty} K(x - y)r(y)\, dy$$

for a kernel, $K\,(x - y)$ that is rather simply determined by the dependence of K_{ij} on distance. When observed values of $r(x)$ and $e(x)$ are substituted in, the resulting estimates of $K\,(x - y)$ are close to those estimated from two-ommatidia experiments. (Attempts to treat the case of a moderately large number of ommatidia, with regard for thresholds, under the assumption that $\tau_{ij} = \tau/K_{ij}$, lead to a nonconvex quadratic programming problem, whose solution is not yet available.)

Problem Formulation: Indirect Measurement of Functioning

We consider now some examples of biological studies in which the mathematical *formulation* challenges the mathematical worker. Here the mathematician finds himself working close to the data with the biologist, using principles of physics and chemistry and biological experience to set out trial relations, test them by analyses or calculations, and then return for refinement. Typically, mathematical difficulties of technique are found to be secondary to formulational difficulties.

The crucial and quantitative aspects of physiology take place deep within the organism. They are often difficult and impossible to measure directly. As a result, especially if human measurements are to become useful in diagnosis, it is frequently necessary to resort to indirect measurement — which must almost inevitably be based on a mathematical model. For the present, these indirect approaches are important in offering physiology what are often its only tools of measurement for important quantities and processes.

Compartment Analysis

One large class of such problems is the analysis of the data taken from radioactive tracer experiments. These problems, called compartment analysis or pool theory, are the counterparts of black-box methods in electric circuit synthesis or input-output analysis in economics. They have been formulated and studied for several years.

In each case, the general approach is the same. The body is divided into "compartments" (which may or may not be physically isolated from one another) and the motion of some substance (an element, an ion, or a compound) through the body analyzed. Two simplifying assumptions are made: (*a*) each compartment is kept continually mixed, so that the concentration of the substance is the same throughout the compartment and (*b*) certain pairs of compartments communicate with each other in such a way that the substance can flow from one to the other at a rate proportional to its concentration in the first compartment. (Such flows may, but need not be, two-way flows, usually at different rates.) When the rate of change with time of the amount of substance in each compartment is set equal to the difference between the total amounts entering and leaving, the model becomes a system of (linear) ordinary differential equations. With enough compartments, such a model can be quite realistic.

If the model is set up and successfully fitted to the data, what is it that is measured indirectly? The rates of transfer from one compartment to another, which are often of major interest, and the "sizes" of the compartments, which as in the case of total blood volume, may also not be susceptible to direct measurement.

Once a model has been set up and adequate data on the time histories of the concentrations in the accessible compartments have been gathered, available computer programs can be used to fit the model's constants by calculation and recalculation until approximately best fitting values are at hand. While the constants to be fitted enter in the simplest way, namely linearly, into the differential equations of the model, they enter nonlinearly into the time histories, thus making the effects of natural biological variability and inevitable measurement fluctuations much more difficult to deal with. As a result, computer programs are not quite ready to be used completely automatically by the nonmathematically trained biological experimenter alone. Questions concerning the appropriate kind and amount of data required to determine the unknowns need to be resolved before the curve-fitting process begins. On the other hand, questions concerning the possible or probable important exchange pathways require the physiologist's insight.

In many such physiological experiments with radioactive tracers, linear models are used because too little is known about the mechanisms to warrant more complicated models. Even in studies of single membrane transport, linear compartment models appear to be still quite useful. There are, however, large classes of chemical kinetic problems that are essentially nonlinear. Much work has been done on enzyme kinetics, and recently much of this has been calculational. Problems involving iron transfer between erythrocytes and other body tissues require nonlinear differential equation models.

An Illustration

In studying how iodine (essentially present as iodide ion) moves through the dog, it is important to consider that part of the system involving the behavior of thyroxin, a major product of the thyroid gland, in which the iodide is bound to a protein. A typical experiment begins by giving a dog a large dose of potassium iodide so that the thyroid gland itself and other body tissues in which unbound iodide may reside are saturated. This, in effect, inactivates the thyroid. Labeled thyroxin is then introduced into the blood. Two activities commence. A portion of the thyroxin is circulated through the body and then diffused into body cells, in which it takes part in metabolic processes. It is then returned to the blood as thyroxin. The other portion is reduced in the blood to unbound iodide. While in the bound form, a fraction of the thyroxin is excreted through the intestines and bowels. The free iodide is passed through the kidney and excreted as urine.

In a typical first trial modeling, circulatory and diffusive processes (by blood and through cell membranes) are taken to occur so rapidly as to approximate complete stirring. If we need only divide the dog into "blood" and "tissues" and if we can neglect the presence of free idoide in the tissues, then, because no iodide is being converted from free to protein-bound, the general picture is that shown in Figure 5.

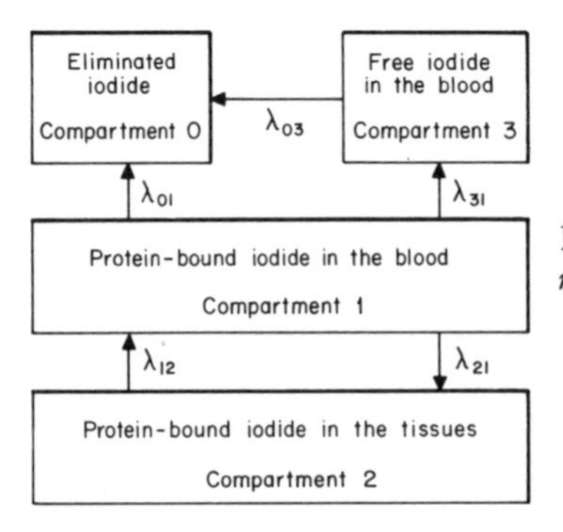

FIGURE 5. *Iodide compartments and interchange rates.*

In an experiment in which only blood measurements are taken, data are obtained as total radiation counts per unit volume in a sample from the blood and, also, the same measure is taken on the precipitated thyroxin in the sample. These are converted to fractions of the total injected radiation count, so that percentage concentrations (of the labeled iodine) are measured at known time intervals. Since the labeled iodide is handled by the dog in exactly the same way as unlabeled iodide, the rate constants shown in Figure 5 apply as well to the labeled iodide as to the total iodide.

If f_i is the amount of labeled iodide in compartment i, K_i is the (effective) size of the compartment, and q_i the concentration of labeled iodide per unit volume, then

$$f_i = K_i q_i,$$

while the differential equations corresponding to Figure 5 are

$$\frac{df_1}{dt} = -(\lambda_{21} + \lambda_{01} + \lambda_{31})f_1 + \lambda_{12}f_2,$$

$$\frac{df_2}{dt} = \lambda_{21}f_1 - \lambda_{12}f_2,$$

$$\frac{df_3}{dt} = \lambda_{31}f_1 - \lambda_{03}f_3.$$

The equations are conveniently written in terms of the f_i since the initial conditions are (at time $t = 0$) $f_1 = 1, f_2 = 0, f_3 = 0$.

Since only q_1 and q_3 are measured, K_2 and q_2 need not be considered separately and are to be only estimated. (K_1 and K_3 need not be the same, although both refer to a "blood" compartment, because free iodide is rapidly "stirred" more widely through the body than is protein-bound iodide.) When these constants are estimated, by use of initial conditions and measured data, they will provide quantitative measures of both transfer rates and compartment volumes. To the extent that the fit to the observed concentrations is satisfactory, it will provide further evidence of the reasonableness of this simple model.

The model discussed has, in fact, been quite successful when used with data obtained by Dr. David Becker and his colleagues at the Cornell Medical College. Transfer rates have been obtained on a series of dogs, and they bear consistent relationships through the series.

Respiration

A rather extensive mathematical formulation has been given for the behavior of chemical elements in the respiratory system. These models again make use of ordinary differential equations, neglecting explicit spatial dependence but introducing time lags and difference equations to account for transport or convection and diffusion times. For a model of cardiopulmonary behavior, gas pressures may be measured in the blood and lungs as data input and output for the models. In this work, the notions of control theory have been introduced.

Some of these systems encompass all areas of blood flow — heart, lungs, brain, and other body regions — whereas others concentrate on special exchange regions. For example, the relation between gas exchange and blood convection in the alveolar region of the lungs has been considered, equations have been written, and a calculational method has been evolved. In this case, since gas content and pressures may be measured easily from a patient, there is reason to believe that high-speed calculation of the model or even on-line computation may eventually produce a clinical tool useful in respiratory malfunction diagnosis.

Education and Collaboration

Although various areas of biology and of medicine have been mentioned in these brief samples, and a reasonable level of mathematical endeavor exhibited in some of them, it is clear there is much to be done. The reasons for mathematicians now to engage in the problems of biology are obvious. The means by which they can do so may not be so simply achieved. There seem to be three main directions that can be chosen:

1. The mathematician may be placed into a laboratory, hospital, or medical school, so that he obtains a close-in view of the experiments and the data and also becomes a target for the local scientist.

2. A biologist may be placed into a mathematics department, where

he can learn techniques of formulation and manipulation, impart his knowledge, and constrain the mathematicians from carrying on unmeasurable or otherwise foolish work;

3. A mathematician may enter into neither of these joint arrangements but he may make use of the abundant literature in each of these fields, bringing mathematics to biology in a more detached fashion.

In all of these approaches, there is the grave problem of trying to create in the mathematician the kind of wisdom and judgment that tells him which of the many problems he is exposed to he should choose to work on. In the established areas of the application of mathematics — fluid and solid mechanics, electromagnetic theory, and so on — this judgment has been nurtured by long years of association and mutual contributions. In biology, it may best be done now by enticing an expert in a particular biological field, who has already used some mathematical methods, to become temporarily a part of a mathematical group and to focus his attention on the formulation and working out of mathematically posed problems in his field. The difficulty is to keep the biologist "wet," to keep him active in his field. In some cases this has been tried successfully. It is dependent partly on superimposing on a biological education an enlightened view of the uses of mathematics. As it is now, most graduates in pure or applied biology understand some of the statistical tools but look upon other model formulations — differential equation systems or Markoff chain probability formulations — as *a posteriori* explications of experimental findings. They have not yet been exposed to the constructive, predictive, or even the experiment-saving virtues of mathematical models. From the mathematician's point of view, at least, it would seem reasonable to begin to expect as much mathematics in the biologist's education as is required in the engineer's.

On the other hand, how shall the mathematician learn biology? For the present, it may be possible to begin treating biological problems the same way fluid mechanics and solid mechanics were handled twenty years ago. The mathematical aspects of gas dynamics, plasticity, wave guides, and the like, were given in graduate courses to people already fairly well trained in mathematics. It would seem possible to introduce a biology option into an applied mathematics graduate curriculum in several areas of biology. It would also be useful to create more postdoctoral opportunities in which graduates in applied mathematics could spend a year or two in a medical or biological institution, working closely with people in a specific field.

It seems reasonable, therefore, to advocate bringing the mathematician to biology and bringing biologists to the mathematicians, both in their education and in their further research work. Past experiences with the third mode of doing biomathematics, the more detached view, have not, in our opinion, proved successful.

BIBLIOGRAPHY

Eugene Wigner, "The Unreasonable Effectiveness of Mathematics in the Natural Sciences," *Commun. on Pure and Applied Mathematics*, *XIII*, 1–14, 1960.

Floyd Ratliff, *Mach Bands*, Holden-Day, New York, 1965.

H. Morowitz and T. Waterman, *Theoretical and Mathematical Biology*, Blaisdell, New York, 1965.

A. L. Hodgkin, *The Conduction of the Nerve Impulse*, the Sherrington Lectures VII, Charles C Thomas, Springfield, Ill., or Liverpool University Press, Liverpool, England, 1964.

D. Noble, "Application of Hodgkin-Huxley Equations to Excitable Tissues," *Physiological Reviews*, *46*, 1–50, January 1966.

BIOGRAPHICAL NOTE

Hirsh Cohen did his graduate work in mathematics at Brown University and, after several years in the academic field, joined the Research Division of IBM, where he is now Director of Mathematical Sciences. After concentrating earlier in his career on problems in fluid mechanics, superconductivity theory, and differential equations, he developed an interest in physiology. Early in 1964 he and some of his colleagues were invited by Cornell Medical School and the Sloan-Kettering Cancer Research Institute to start looking into mathematical problems in medicine and biology. They formed the nucleus of a permanent staff with responsibilities in teaching, research, computing, and other mathematical services. This group at Cornell and Sloan-Kettering is beginning to train students for the Ph.D. in the new field of biomathematics, besides collaborating with experimental biologists and clinical medicine people. It is among the first of several biomathematical departments now being organized.

Laplace, a famous contributor to this branch of mathematics, once wrote that "the most important questions of life are, for the most part, really only problems of probability." At one time mathematicians had serious misgivings about probability theory; the logical foundations seemed flimsy. But after they learned how to resolve some confusing paradoxes, probability theory developed rapidly until now it is one of the chief areas of contact between mathematics and the world as a whole. This essay describes some of the important ground rules for using probability theory and discusses how it can be applied to help reconcile man with a universe where uncertainty is rampant.

Probability

Mark Kac

The impact of probability theory on science and technology has in recent years been well-nigh miraculous; its role as a weapon of rational approach toward an amazingly wide variety of problems has been, to say the least, unique. All this is now well known and well recognized.

Probability theory occupies a unique position among mathematical disciplines because it has not yet grown sufficiently old to have severed its natural ties with problems outside of mathematics proper, while at the same time it has achieved such maturity of techniques and concepts it begins to influence other branches of mathematics. Thus probability theory provides a remarkably congenial domicile for interplay of ideas and for new and often unexpected connections.

At the most elementary level, probability theory is concerned with calculating or estimating odds of events that can occur only in a finite number of ways. A typical example would be to calculate the probability that a poker hand will be a flush. Assuming all poker hands to be "equally likely," a natural definition of this probability is the ratio:

$$P = \frac{\text{number of poker hands that are flushes}}{\text{total number of poker hands}}.$$

This is a special case of the definition of probability given by the father of probability theory, Pierre Simon, Marquis de Laplace (1749–1827).

Laplace defined the probability $p(E)$ of an event E as the ratio of the number $n(E)$ of ways in which the event can be realized to the total number N of possible outcomes:

$$P(E) = \frac{n(E)}{N},$$

provided that all possible outcomes can be assumed equally likely. With this definition, elementary probability theory is reduced to combinatorial analysis, an important (and often very difficult) branch of mathematics devoted to what may be termed "counting without counting." Typical examples of problems in combinatorial analysis are: In how many ways can ten couples be seated around a table so that no wife sits next to her husband? or In how many ways can ten letters be placed in ten envelopes so that no letter is placed in the proper envelope?

We can rephrase the latter example in probabilistic terms by asking for the probability that a secretary who decides to place ten letters in envelopes "at random" will miss every single one. It may be surprising to learn that this probability is very close to 1/2.718 . . . and is hence quite high (2.718 . . . is a famous number called e to which several references will be made in the sequel).

Arithmetic of Choices

A very important problem in combinatorics is to find the number C_n^k of ways of choosing k objects out of n. When n and k are small, we can solve this problem by direct enumeration. For example, there are six ways of picking two out of four letters A, B, C, D, namely AB, AC, AD, BC, BD, and CD. Thus $C_4^2 = 6$.

There is also a simple general formula for C_n^k, that is,

$$C_n^k = \frac{n(n-1)(n-2)\cdots(n-k+1)}{1\cdot 2\cdot 3\cdots k},$$

and the numbers C_n^k can be conveniently displayed in Figure 1, the so-called Pascal triangle, where the entries in the nth row represent

$$C_n^0 = 1, \qquad C_n^1 = n, \qquad C_n^2, \ldots, C_N^{n-1} = n - 1, \qquad C_n^n = 1.$$

That there is much more to probability theory than mere counting became apparent with the discovery of so-called laws of large numbers. Suppose I toss ten (10) coins: What is the probability that exactly four (4) heads will show? An outcome of ten tosses can be recorded symbolically as a sequence (of length ten), for example,

HHTTTHTTHT,

where H stands for "heads" and T for "tails." There are $C_{10}^4 = 210$ such sequences containing exactly 4 H's (for there are C_{10}^4 ways of choosing from the 10 positions in the sequence the 4 that are to be occupied by H's). We get the total number of sequences of length ten by adding up

FIGURE 1. *Pascal's triangle, an aid to calculating probabilities, is made up of the coefficients of the binomial expansion. Each number is the sum of the two numbers immediately above it. Some of its characteristics and applications are discussed in the text.*

all the entries in the tenth row of the Pascal triangle, which comes to $1024 = 2^{10}$. Now, if the coins are "honest," in the sense that all sequences of length ten are equally likely, the desired probability is

$$\frac{210}{1024} = \frac{C_{10}^4}{2^{10}}.$$

The fact that the sum of all entries in the tenth row of the Pascal triangle added up to 2^{10} is not an accident. In general, the sum of all entries in the nth row is 2^n, and thus the probability that, in tossing n "honest" coins, exactly k "heads" will show is $C_n^k/2^n$.

Now let the probabilities that exactly 0, 1, 2, 3, . . . , 10 heads will show in 10 tosses be displayed graphically, as shown in Figure 2. The bases of the eleven rectangles are all equal to one, while the heights are the probabilities. Thus, the height of the rectangle whose base is the interval between 4 and 5 is the probability that in ten tosses exactly four heads will show, that is, 210/1024 or roughly, 21.

The graph is symmetric with respect to the midpoint (5, 5) and is rather wide and low (the maximal height is 252/1024 or roughly .25). If, instead of ten tosses, we were to take ten thousand (10,000) tosses, the corresponding graph would be centered around 5,000.5; it would have width 10,001; and its maximum height would be, to a very good approximation, $1/100\sqrt{\pi} = .0056$. If, however, we place ourselves at 5,000.5 and increase all heights by the factor $100/\sqrt{2} = 70.70...$ while at the same time shrinking the basis by the same factor $(100/\sqrt{2})$, the "rectangle

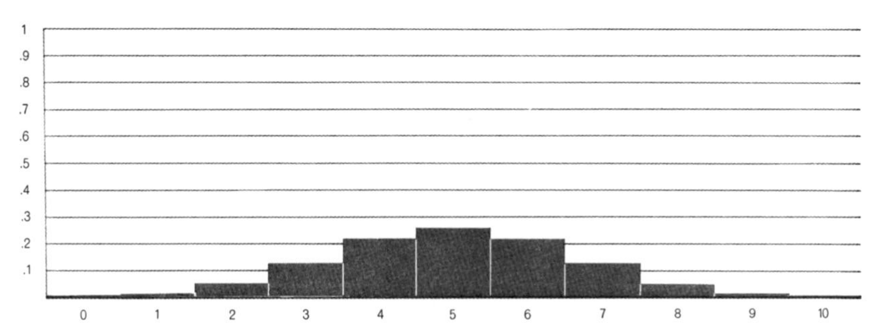

FIGURE 2. *Probability of heads in 10 tosses of a coin produces a histogram that is reminiscent of the normal distribution. As the 10th row of Pascal's triangle shows, there are 210 possible sequences containing exactly four heads out of a total of 1,024 possible sequences of heads in 10 tosses. Thus the chance of four heads is about 21 percent. Roughly the chances of zero, one, two, three, four, and five heads in 10 tosses* (horizontal scale) *are respectively .001, .01, .045, .12, .21, and .25* (vertical scale). *The probabilities and the bars representing them occur in descending order for six through 10 heads in 10 tosses. On the same scale a histogram for 10,000 tosses would be much wider and lower, and would have to be rescaled to bring out its relation to the normal curve.*

tops" would become all but indistinguishable from the continuous curve described by the equation

$$y = \frac{1}{\sqrt{2\pi}} e^{-x^2/2},$$

where $e = 2.718...$ is the celebrated number John Napier took as the base for his "natural logarithms." (If a bank were foolish enough to offer 100 percent interest annually and, in addition, agree to compound it continuously — not just daily, hourly, or even every second, but every instant — one dollar would grow to exactly e dollars at the end of a year.)

The larger the number n of tosses, the better will the rectangular graph approximate the continuous curve. Of course, in case of n tosses we must stand at $(n + 1)/2$, magnify the heights by the factor $\sqrt{n}/2$ and shrink the bases by the same factor. The appearance of this remarkable regularity for large n is an example of what is technically known as a law of large numbers.

The rectangular graphs I have been describing are called *histograms*, and they are widely used in statistics as a device for pictorial presentation of data. Consider a population of women whose heights have been measured and recorded. We can display these data by first agreeing on some basic interval size (for example, 1 inch) and then drawing rectangles whose bases are 1 inch long and whose altitudes are equal to the number (or percentage) of women whose heights fall within these intervals. For instance, the height of the rectangle whose base is the interval between 60 and 61 inches will be 100 if this happens to be the number of women in the population whose heights fall within the range between 60 and

61 inches. If the size of the interval is chosen too small, the histogram will be quite irregular, and, if it is too large, the histogram will be dull and featureless. By our choosing the interval size just right, the rectangle tops may again be well approximated by a smooth curve. All this is illustrated in Figure 3.

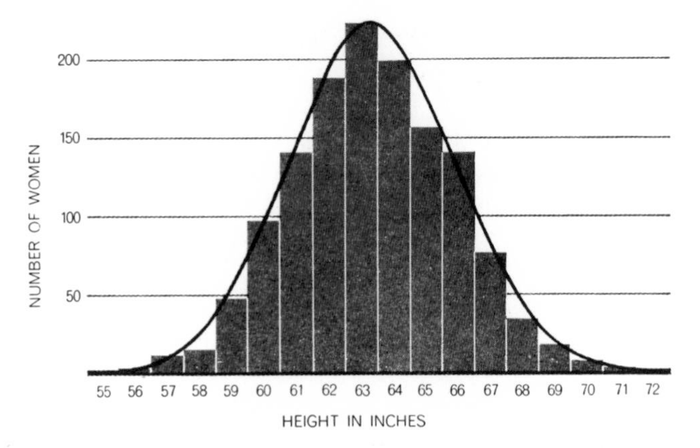

FIGURE 3. *Heights of women produce a histogram to which the normal-distribution curve can be fitted. There were 1,375 women in this sample population. The bell-shaped curve conforms to many other empirical distributions found in the physical and biological worlds.*

The graph in Figure 2 is the histogram of the number of heads in 10 tosses. As we have seen for a large number n of tosses, the histogram of the number of heads when properly centered and scaled is almost identical with the continuous curve

$$y = \frac{1}{\sqrt{2\pi}} e^{-(x^2/2)},$$

known as the *normal curve.*

This curve is among the most celebrated in science. It appears over and over again as an empirical law, serving to describe the variability of parameters in large populations (heights of men and women, sizes of peas, weights of newborn babies, and so on) and in a more disciplined and rational way, it turns out as the answer to many important problems of physics (distribution of velocities in a gas, distribution of displacements of Brownian particles, and so forth). That there was a simple mathematical model, based on coin tossing, that in a rigorous way led to this remarkable curve was, of course, highly suggestive.

Brownian Motion

Before we go on, let us pause to discuss briefly how our considerations concerning coin tossing can be applied to the theory of Brownian motion.

In 1828, the English botanist Brown observed under a microscope that small particles of India ink suspended in a liquid perform strange erratic movements that now bear his name. The first theories of this curious phenomenon were given in 1905 by Einstein and Smoluchowski. In his approach to the problem, Smoluchowski used the *random walk* picture, which is as follows.

Imagine a particle jumping along a straight line in steps of length d cm, each step taking τ sec. And imagine that, each time the particle is about to jump it tosses an imaginary "honest" coin: On "heads" it jumps to the right, on "tails" it jumps to the left.

The motion of the particle depends, of course, on the vagaries of luck, and its position cannot be predicted with certainty. We can, however, ask for the probability that at some time the particle will be found in a prescribed interval. This can be translated at once into a question concerning the number of heads in a certain number of tosses of a coin, and it should not therefore be surprising that the answer involves the normal curve. The answer also involves (in a simple way) the ratio $d^2/2\tau$ (the so-called diffusion constant), and this in turn involves the Avogadro number, that is, the universal number of molecules in a mole of every substance.

By watching a large number of Brownian particles under a microscope for the same length of time and measuring their horizontal (or vertical) displacements, we obtain the kind of "raw" data that are not unlike heights of women. (See Figure 4.) If we choose an interval size appropriately (it will depend on the duration of observations and on the diffusion constant $d^2/2\tau$), we can bring the histogram of the horizontal (or vertical) displacements into excellent agreement with the normal curve.

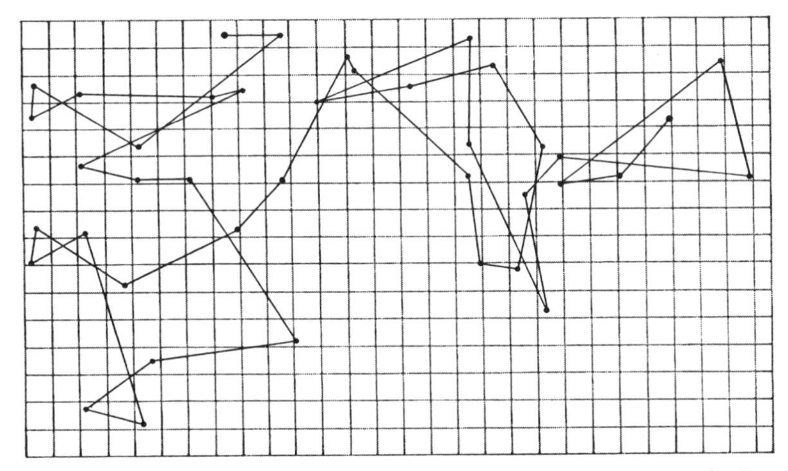

FIGURE 4. *Brownian path is taken by a particle being "kicked around" by molecules of a surrounding liquid or gas. A stochastic process (it varies continuously with time), Brownian motion can be analyzed and modeled by probabilistic techniques.*

When this is done, the empirically determined interval size can be equated to the one predicted theoretically (which depends on the time of observation and on the diffusion constant); thus, a way is paved to a determination of the Avogadro number from Brownian motion experiments.

Actual experiments of this kind were performed by Jean Perrin during the early years of this century. For this he was later awarded the Nobel Prize in Physics. The successful determination of the Avogadro number from Brownian motion experiments was a landmark in the history of science. It led to a belated but universal acceptance of atomistic theories and made possible other great advances in exact sciences.

How prophetic, in retrospect, were the words of Laplace, when he said, "It is remarkable that a science which began with the consideration of games of chance should have become the most important object of human knowledge."

In spite of Laplace, probability theory all but disappeared as a mathematical discipline during the nineteenth century and the first two decades of the twentieth. Even the spectacular applications to physics, beginning with Maxwell and Boltzmann and culminating in the theory of Brownian motion of Einstein and Smoluchowski did little to change the indifferent attitude with which the mathematical community of the day treated what Laplace considered "the most important object of human knowledge."

Poincaré and Hilbert, the two greatest mathematicians of the late nineteenth and early twentieth century, tried to revive interest in probability theory. Poincaré lectured on the subject and his lectures appeared in book form. It was a highly original book (everything Poincaré did was highly original), but it had remarkably little influence. Hilbert included axiomatization of probability theory (lumping it with other branches of "applied mathematics") in his famous list of problems (presented to the International Congress of Mathematicians held in Paris in 1900), whose solutions he considered to be of paramount importance to the future development of mathematics. But somehow no one took up the challenge until much later.

It was only in Russia that a remarkable genius, P. L. Chebyshev (1821–1894), carried on the great work of Laplace and brought the subject to the threshold of modern times. His student, A. A. Markov (1856–1922) made even more remarkable contributions and started many developments, some of which are still vigorously pursued today (for example the theory of so-called Markov chains, which we shall discuss a little later). Chebyshev and Markov are, without doubt, responsible for the great strength of probability theory in the Soviet Union today.

The Plague of Paradoxes

It is futile to dwell at length upon the reasons for the neglect of probability theory. There is, however, no doubt that looseness of logical foundations combined with rapidly rising standards of rigor had a great deal to do with it. There were, first of all, logical objections to Laplace's definition of probability. For this definition, as we have seen, we need the notion of *equally likely outcomes*, and this already involves the notion of probability. This brings us close to a circular definition, which is absolutely taboo in mathematics.

But this was not the worst. The gradually developing field of so-called geometric probabilities was being plagued by seeming paradoxes and other difficulties. The most celebrated of all problems in geometrical probabilities is the Buffon needle problem:

If a needle of length l inches is thrown "at random" on a floor made out of planks of width d inches, what is the probability that the needle will intersect a crack between planks? Assuming that the length l of the needle (l) to be less than the width (d) of the planks the usual solution is as follows.

First we may assume that the center of the needle falls within a particular plank. Thus the position of the needle is completely determined by x, the distance from the center of the needle to the lower crack, and by θ, the angle from the lower crack to the needle measured counterclockwise. Since x lies between 0 and d and θ between 0 and π (in all serious mathematics angles are measured in *radians* and in these units 180° is π, 90° is $\pi/2$, 60° is $\pi/3$, and so on), a throw of a needle corresponds uniquely to a point in a rectangle whose base is π and whose altitude is d (width of the plank). The correspondence between the position of the needle and a point in the rectangle is illustrated in Figure 5. The points of the rectangle represent all possible outcomes of our experiment. What remains is to describe the set of points that correspond to outcomes resulting in the needle's intersecting one of the cracks. To do this, one needs simple trigonometry, and it turns out that the set in question consists of the two shaded areas.

So far, everything is incontrovertible and beyond reproach. But now we take a leap and assume that the probability is the ratio of the shaded areas to the total area of the rectangle. The area of shaded portions can be calculated by use of elementary calculus, and the ratio of the areas turns out to be $2l/\pi d$. But while there can be no objection to choosing x and θ as parameters defining the position of the needle, there surely is no compelling reason to treat all points in our rectangle as being in a sense "equally likely," an assumption that forces us into defining probability as the ratio of areas.

The degree of arbitrariness involved in problems of geometric probabilities was dramatized by the French mathematician L. F. Bertrand, who in 1899 discussed the following problem: What is the probability that a "random" chord in a circle will be shorter than a side of an inscribed equilateral triangle?

We can begin by agreeing that, since one end of the chord can be

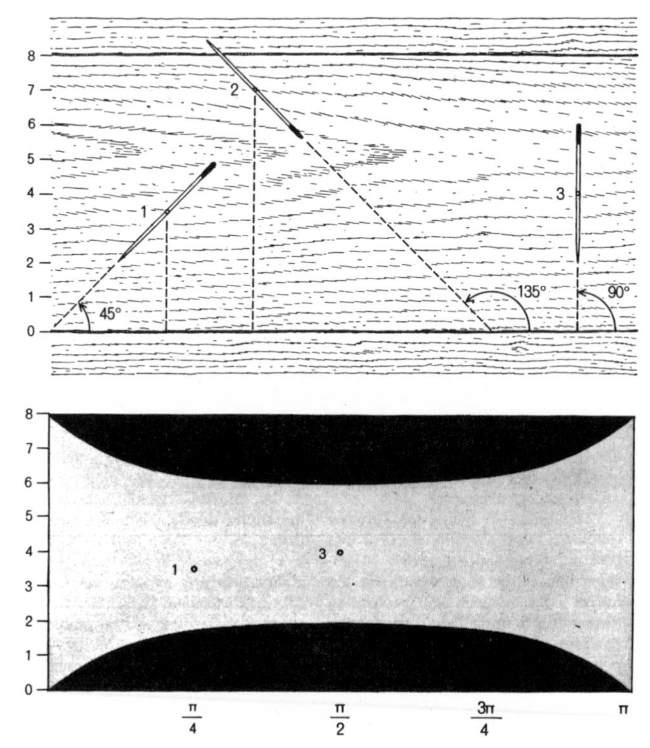

FIGURE 5. *Buffon needle problem involves the probability that a needle shorter than the width of a plank will fall across the crack between two planks. Here each needle is half as long as the plank is wide. The eight units used to measure the plank represent inches. Abstract diagram also shows positions of the three needles. The horizontal scale represents the angle of each needle with respect to the bottom edge of the plank. The angle is given in terms of π, which is defined as 180 degrees. The vertical scale is the width of the plank in inches. The three dots are the center points of the needles. Called the "sample space," the rectangle represents all the possible positions in which a needle can fall. The darker areas cover all the positions in which a needle lies across a crack.*

chosen arbitrarily, we may as well fix it at P (see Figure 6) and leave to "chance" only the choice of the other end Q. This is equivalent to choosing "at random" the central angle θ, and, if all such angles between 0 and 2π are assumed "equiprobable," the answer to the problem turns out to be 2/3.

If, on the other hand, we leave to "chance" the choice of the midpoint M of the chord, then for the chord to be shorter than a side of an inscribed equilateral triangle, M will have to fall outside the circle of radius half of the original one, that is, in the shaded band in Figure 7. Assuming all points in the circle to be "equiprobable," the desired probability is the ratio of the shaded area to the total area of the circle, which is clearly 3/4.

This "paradox" is, of course, no paradox at all, for anyone can define

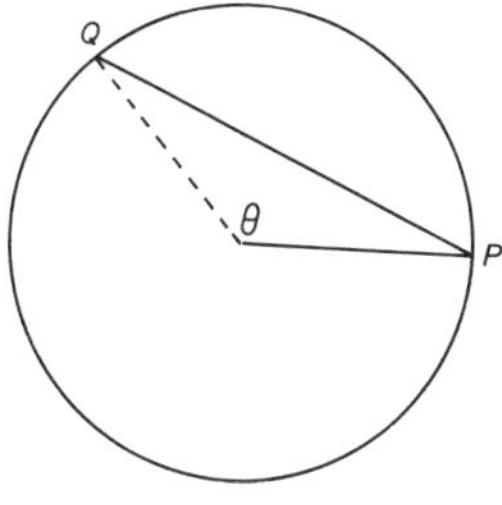

FIGURE 6

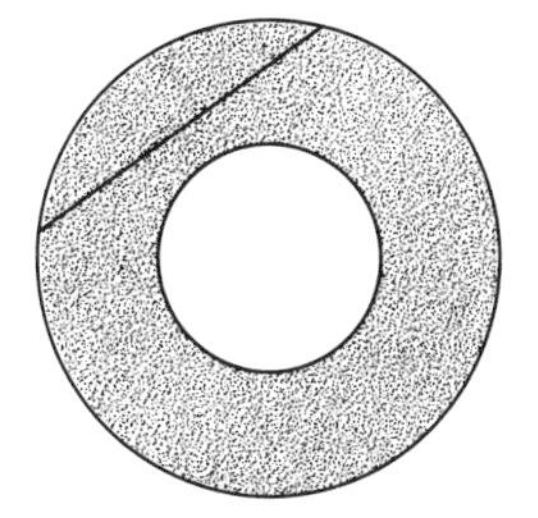

FIGURE 7

"at random" in any old way he pleases. But it is not too surprising that a subject so loose as to allow such latitude of interpretation was looked upon with grave suspicion at the time when rigor and precision were gaining the upper hand in mathematics.

Meeting at Random

How do we look upon all this now? I can explain this best by considering an example. Suppose that two friends living in different and somewhat remote suburbs of New York would like to meet in front of the Public Library. Transportation being what it is, they can be sure of getting there only between 12:00 noon and 1:00 P.M., the exact times of arrival being subject to chance. Not wishing to waste too much of each other's time, they further agree to wait 10 minutes ($\frac{1}{6}$ of an hour) from the time of arrival or till the end of the hour, whichever is sooner. What is the probability that they will actually meet?

Let the times of arrival of the two friends be denoted by x and y. All we know in advance is that both x and y lie between 0 (noon) and 1 (1:00 P.M.). We can thus associate with each pair of times of arrival a point (x, y) in the "unit square" whose coordinates are x (abscissa) and y (ordinate). The "unit square," like the "θ, d-rectangle" of the Buffon needle problem, represents all possible outcomes. For a meeting to take place, the difference between the times of arrival must be between $-\frac{1}{6}$ and $+\frac{1}{6}$, and this defines the shaded region in Figure 8. If again the probability is taken to be the ratio of the areas, the answer to our problem is $\frac{11}{36}$.

Suppose now there were three friends involved (all living in different suburbs), and we wanted the probability that at least two would meet. Instead of a square, we will now have a cube, and the shaded region will be much more complicated, besides being three-dimensional.

If there are more than three people, we can no longer appeal to pictures, but you should nevertheless be convinced that the problem can be handled by extending the usual theory of areas and volumes to many-dimensional spaces. This indeed can be done, although in many cases the calculation of multidimensional volumes can be extremely difficult. An example is the following problem.

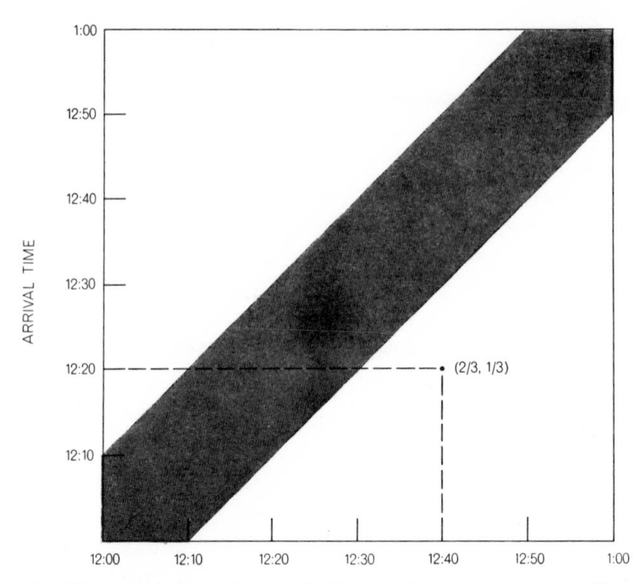

FIGURE 8. *Possible arrival times of two suburbanites planning to meet at a library between 12:00 noon and 1:00 P.M. can be plotted. Arrival times for one person are on the vertical scale, for the other on the horizontal scale. Shaded area covers region corresponding to a meeting. In order to meet they must arrive at library within 10 minutes of each other. As can be seen, if one arrives at 12:20 (a third of the way up) and the other at 12:40 (two-thirds of the way across), they will not meet. The* $(\frac{1}{3}, \frac{2}{3})$ *point falls outside the shaded region.*

What is the probability that when N points are chosen at random in a cube of side L no two of them will be closer than a prescribed distance δ? (Since each point has three coordinates, the choice of N points is tantamount to choosing *one* point in a $3N$-dimensional cube of side L. "At random" means here that all points in the $3N$-dimensional cube are equiprobable.)

This is an important unsolved problem in statistical mechanics. Its solution would shed much light on the theory of changes of state of matter (for example, from solid to liquid, as in melting).

Let us go back to our two suburban friends trying to meet in front of the Public Library. The assumption that all points in the square are "equiprobable" hides, in reality, two vastly different assumptions. The first assumption concerns the individual times of arrival. What we assume here is that the probability that x (or y) will fall within a time interval of duration t (less than one hour) is simply equal to t. In technical terminology, we say that both x and y are uniformly distributed in the interval from 0 (noon) to 1 (P.M.). The second assumption concerns the interaction between the times of arrival of the two persons involved. It states that the probability the first person will arrive in a time interval l_1 and the second in a time interval l_2 is the product of the two individual probabilities.

The first assumption has something to do with how frequently the trains run and with how reliable they are about adhering to their schedules. If there is only one train scheduled to arrive in New York between 12:00 noon and 1:00 P.M. and if the scheduled time of arrival is 12:20 P.M., then, allowing 15 minutes for the walk from Grand Central to the Public Library, one is much more likely to arrive at the ultimate destination after 12:35 P.M. than before. Under these circumstances, the first assumption, though mathematically allowable, would be completely unrealistic. If, on the other hand, there are many trains (say ten or so) scheduled to arrive roughly every six minutes and the crews are a bit sloppy about keeping on time, the first assumption (uniform distribution of times of arrival) may not be bad at all as an attempt to describe reality.

To describe different kinds of unpredictability of times of arrival, one can make use of the fruitful concept of the *probability density curve*. Specialized to our case, it is a curve over the interval from 0 to 1 of total area equal to unity (corresponding to certainty of arrival between noon and 1:00 P.M.) and such that the area over a time interval l represents the probability of arriving during that time interval. In the case of the sole 12:20 train, the probability density curve may look like that in Figure 9

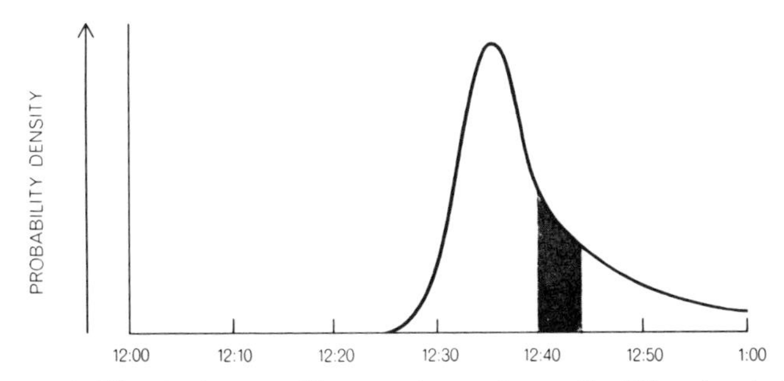

FIGURE 9. *Probability-density curve illustrates degree of unpredictability of arrival time if there is only one train, coming in at 12:20. Most likely meeting time at library is 12:35. Area of shaded portion represents probability of arrival between 12:40 and 12:44* P.M.

while, in the case of ten trains, the probability density curve is well approximated by the dull straight line (Figure 10) denoting the case where all times of arrival are "equally likely."

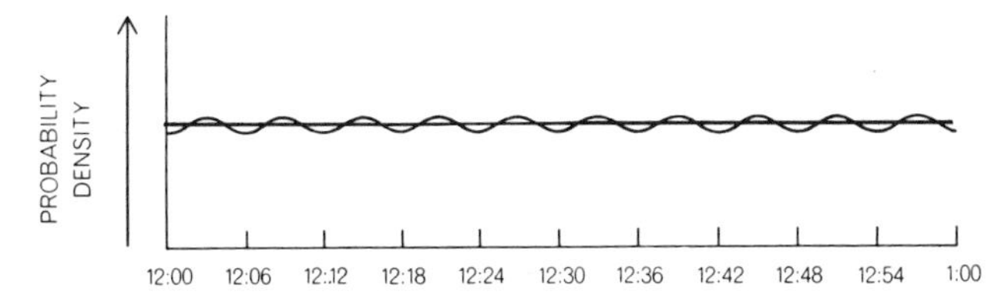

FIGURE 10. *Trains arriving every six minutes give density curve shown by wavy line. The straight line is the "curve" in case all the arrival times are equally likely.*

The second assumption (probability that the time of arrival of the first person falls within an interval l_1 and the time of arrival of the second within an interval l_2 is the product of the two individual probabilities) is a translation into mathematical language of the belief that the times of arrival are independent of each other. The notion of independence is of central importance in probability theory, and we pause to consider why it is reflected in the rule of multiplication of probabilities and to discuss some related questions.

Compounding Problems

Let us introduce a few notions and some terminology. With each experiment, we associate the set of all its possible outcomes and call it the *sample space*. For example, in the experiment of tossing 10 coins (or tossing 1 coin 10 times), the sample space is simply the set of all possible "words" of length 10, each letter being either H or T. A typical word is

HHTTTHTHHT,

and the whole sample space consists of 2^{10} words (or *points*, as elements of any sample space are often called.

For the Buffon needle problem, the sample space can be taken to be the θ, d-rectangle; for the problem of n suburbanites wanting to meet between noon and 1:00 P.M. in front of the Public Library, the sample space is the "n-dimensional unit cube."

Events now become identified with subsets of the sample space. For example, the event that in 10 tosses of a coin exactly 4 heads will show corresponds to the set of words of length 10 containing exactly 4 letters H. The event that the needle will intersect a crack in the floor corresponds to the set consisting of the two shaded regions shown in Figure 5, and so on.

In very general terms, probability theory, as a mathematical discipline, is concerned with the following problem: Given a collection of "elementary" events whose probabilities are somehow known (or postulated), how do we go about calculating probabilities of more complex events?

It is customary to assign probability 1 to the whole sample space and to use the following two rules:

I. If E_1, E_2, . . . are mutually exclusive events (that is, the occurrence of one precludes the occurrence of any other) whose probabilities $p(E_1)$, $p(E_2)$, . . . are somehow known, then the probability of the event (E_1 or E_2 or E_3, . . .) is the sum of the individual probabilities. Symbolically,

$$p(E_1 \text{ or } E_2 \text{ or } E_3 \text{ or } \cdots) = p(E_1) + p(E_2) + \cdots .$$

II. If the probability of events E and F are $p(E)$ and $p(F)$, respectively, and if F implies E (that is, if F happens, we can be sure that E will happen too), then the probability of the event (E but not F) is $p(E) - p(F)$.

These rules are axioms and are hence taken for granted. They can nevertheless be heuristically justified (but not proved) as follows.

We intuitively feel that in repeated experiments (for example repeated throws of a needle) the frequency with which an event E occurs should approximate the mathematical probability $p(E)$ of the event, the approximation becoming better as the number of experiments increases.

If in such n trials the event E_1 occurs n_1 times, E_2 occurs n_2 times, and so on, then the event (E_1 or E_2 or E_3 or $\cdots$) occurs

$$n_1 + n_2 + n_3 + \cdots$$

times because no overlap in occurrences is possible (recall that the events $E_1, E_2, \cdots$ are mutually exclusive). Thus, the frequency of the event (E_1 or E_2 or E_3 or $\cdots$) is

$$\frac{n_1 + n_2 + n_3 + \cdots}{n} = \frac{n_1}{n} + \frac{n_2}{n} + \frac{n_3}{n} + \cdots,$$

the sum of the frequencies of the events $E_1, E_2, E_3, \ldots$. Consequently, if the theory is to have any connection with reality, we must require the first rule to hold. Similarly, we justify (but do not prove) the second rule.

To make plausible the rule of multiplication of probabilities of independent events, we can use the same intuitive feeling that frequency should approximate the mathematical probability.

Suppose that in n trials the event E has occurred n_E times and the event F, n_F times. Further, let n_{EF} denote the number of times the events E and F have occurred simultaneously. If E and F are independent, we feel that the frequency with which F occurs among the occurrences of E should be roughly the same as the over-all frequency with which F occurs, for it is appealing to interpret independence to mean that the knowledge that E has occurred should be of no help in predicting the occurrence or non-occurrence of F.

Thus

$$\frac{n_{EF}}{n_E} \quad \text{should be roughly equal to} \quad \frac{n_F}{n}$$

or

$$\frac{n_{EF}}{n} \quad \text{should be roughly equal to} \quad \frac{n_E}{n} \times \frac{n_F}{n}.$$

From the purely mathematical point of view, we forget all about frequencies and heuristic justifications and define events $E_1, E_2, \ldots$ to be independent to mean that the probability of simultaneous occurrence of $E_1, E_2, \ldots$ is the product of the individual probabilities; symbolically

$$p(E_1 \ \underline{\text{and}}\ E_2 \ \underline{\text{and}}\ E_3 \ \underline{\text{and}}\ \cdots) = p(E_1)\, p(E_2)\, p(E_3) \cdots.$$

What is now the status of Laplace's "equally likely" outcomes? Suppose I toss two dice. There are altogether 36 possible outcomes, and assigning to them equal weights (or probabilities) is once again a combination of two quite different assumptions.

1. Each die is "honest," that is, each side of the cube has probability $\frac{1}{6}$.
2. The dice are independent.

The first assumption can be violated by "loading" one or both of the dice. To violate the second, one must somehow "couple" the dice. We can imagine, for instance, the dice to be connected by a string which would not interfere with individual faces being equiprobable but which would tend to make the pairs (1, 1), (2, 2), . . . , (6, 6) somewhat more probable than the others, thus destroying independence. If there is no coupling at all (for example, the dice are tossed on different planets), we may feel safe in using the rule of multiplication of probabilities.

If we compare the problem of tossing two dice with the problem of the two suburbanites, we see how much alike they are. The main difference is that in the first problem the sample space is finite (it consists of 36 "points"), while in the second, being a square, it is a continuum. Also, in the second case, "loading" corresponds to violating the uniform distribution of times of arrival while "coupling" may be achieved by telephoning prior to choosing a train.

The rules (axioms) for calculating probabilities of complex events are identical with those used for calculating areas and volumes in geometry. In plane geometry, for example, starting with the assumption that the area of a square is the square of its side, we proceed by judicious use of the two axioms to determine areas of polygonal figures. It should, of course, be clear that in applying the axioms the word "event" should be replaced by the word "set," and the word "probability," by the word "area." (In geometry, however, we cannot insist, for good reasons, that the area of the whole plane be one, corresponding to the universal acceptance of the convention that the probability associated with the whole sample space be 1.)

To be able to assign areas to sets bounded by curves (for example, area of a circle), we must allow in Axiom I infinitely many sets $E_1, E_2, \ldots$ (complete additivity, as it is called), and, similarly, we allow infinitely many events when Axiom I is used in probability theory.

It turns out that by postulating complete additivity, we increase vastly the collection of sets to which we may assign unambiguously an area (or *measure*, as it is referred to once we leave the realm of elementary sets). The Greeks already had difficulties with circles and parabolas; they would be appalled if they knew to what complex sets we can now assign measures without so much as a shudder.

The problem of assigning measures to sets has never ceased to interest mathematicians, and during the early years of our century a new and fruitful branch moved into the forefront of mathematics. It is called *measure theory*, and it owes its greatest spurt to Emile Borel and to Henri Lebesgue.

Coin Tossing, Continued

During most of the nineteenth century and during the first two decades of our century, when probability theory was neglected and all but forgotten, the few mathematicians who kept the great tradition of Laplace alive concentrated mainly on extending the remarkable connection between the numbers C_n^k and the normal curve. The culmination of this investigation became embodied in what is now known as the *central limit theorem*. As is often the case with mathematics, the struggle with an important problem produced valuable fringe benefits. A great deal of mathematics, outside of probability theory, owes its growth and development, directly or indirectly, to the central limit theorem. While the trend was toward greater and greater generality, simple coin tossing never ceased to be a source of new problems and new lines of inquiry.

An example of a recent discovery (which would have been well within the scope of Laplace and his contemporaries) that stimulated much interesting and important further work is the so-called "arcsine law." Suppose a player bets one dollar each time an honest coin is tossed, winning on "heads" and losing on "tails." In a long sequence of trials, we can ask for the probability that the proportion of time during which he is ahead will be less than some prescribed fraction f. The answer (which explains the name of the law) is $(2/\pi)$ arcsin $\sqrt{f}$, where arcsin $\sqrt{f}$ means the angle (in radians) whose sine is $\sqrt{f}$. For those whose trigonometry is (at the moment) a bit shaky, the following two numerical illustrations should suffice. The probability of being ahead less than 25 percent of the time turns out to be $\frac{1}{3}$. But more dramatically, if the "honest" coin is tossed every second for a whole year, then with probability as high as $\frac{1}{20}$ the player will be ahead only during less than 13.5 hours! A player afflicted with such seemingly bad luck would have a hard time believing that the coin was not loaded against him.

Is Chaos Inevitable?

Much has been said and written about philosophical implications of probability theory. This is too vast a subject to be taken up here. It is appropriate, however, to include a brief discussion of the role probability theory played in resolving conceptual difficulties involved in reconciling the mechanistic and the thermodynamic view on the structure and behavior of matter.

If two containers, one (A) containing gas under pressure p and the other (B) evacuated, are suddenly connected, thermodynamics predicts a unidirectional flow of gas from A to B, the pressure tending exponentially toward equalization. This is a consequence of the famed Second Law of

Thermodynamics, which, in its most pessimistic form, predicts ultimate equalization of everything, leading to what Clausius called *Wärmetod* (heat death).

The mechanistic (kinetic) view, picturing matter as composed of particles (molecules) obeying the laws of dynamics, leads to an entirely different picture. The molecules of the gas, bouncing against one another and against the walls in what appears to be "random" fashion, will surely not produce an absolutely unidirectional flow from A to B.

As a matter of fact, by a theorem of Poincaré a dynamical system, like the one under consideration, will (unless it starts from certain exceptional states that can safely be neglected) eventually come back, arbitrarily near its initial state. This "quasi-periodic" behavior of dynamical systems contrasts sharply with the monotonic (and monotonous) trend to equalization implied by the Second Law.

To clarify the issues involved, Paul and Tatiana Ehrenfest proposed in 1907 a simple and beautiful probabilistic model. Consider two boxes, one of which (A) contains initially a large number N of numbered balls. We now play the following game, shown in Figure 11. We pick "at

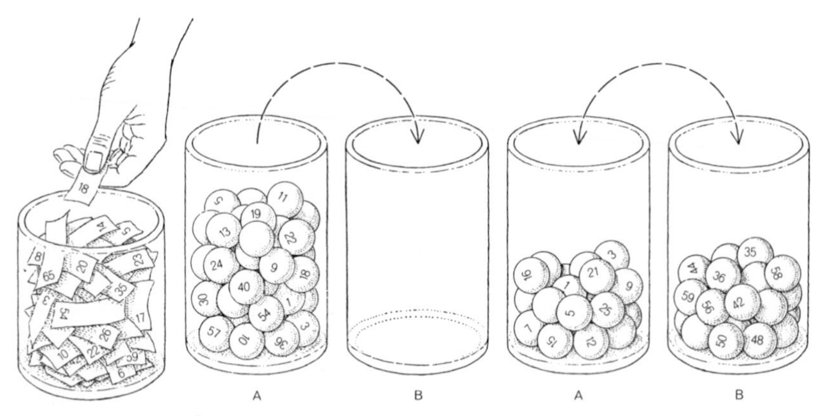

FIGURE 11. *Ehrenfest model for illustrating a Markov chain involves a game in which balls are moved from one container to another according to numbers drawn at random from a third container* (left). *As long as there are many more balls in container* A *than in container* B, *the flow of balls will be strongly from* A *to* B. *The probability of finding in* A *the ball with the drawn number changes in a way that depends on past drawings. This form of dependence of probability on past events is called a Markov chain.*

random" a number between 1 and N and move the ball of that number from whichever box it is in to the other one. (The first move will thus result in moving some ball from A to B.) The process is repeated many times, consecutive drawings being independent with all numbers from 1 to N equally likely.

It is intuitively clear that, as long as there are many more balls in A

than in B, the probability of moving from A to B will be correspondingly greater. We can thus expect a sort of preponderant flow from A to B.

Although the drawings are independent, the quantities of balls in A at consecutive instances are not. They exhibit a kind of dependence first studied by Markov and called a Markov chain. Many Markov chains are simple enough to be amenable to an exact, quantitative discussion, and this is fortunately true of the Ehrenfest model.

We find that the average number of balls in A decreases exponentially to $N/2$, in complete agreement with the thermodynamic prediction. We also find that, with probability equal to unity, the model will eventually return to its initial state (that is, all balls in A). This is clearly the counterpart of Poincaré's theorem about dynamical systems.

It now becomes evident that there is no real contradiction between the Second Law and the inherently quasi-periodic behavior of dynamic systems, provided we are willing to give up the absolute dogmatism of the Second Law and allow a more flexible interpretation, based on probability theory. All this is reinforced if one calculates how long, on the average, he would have to wait for the return of the initial state in the Ehrenfest model. The answer is 2^N steps, which even for moderate N (say of the order of 100) is staggeringly large. If we observe all around us irreversible (unidirectional) behavior in nature, it is simply because our life span is so pitifully short compared with those enormous times of return.

With modern high-speed computers, it is easy to play the Ehrenfest "game." Experiments were performed on the IBM 7090 with $N = 2^{14} =$ 16,384 "balls," each run consisting of 200,000 drawings (it takes less than 2 minutes). The number of balls in A was recorded every 1,000 drawings, and one of the resulting graphs is shown in Figure. 12.

As one can see, at first the number of balls in A falls off along a nearly perfectly exponential curve. But near equilibrium the graph gets a bit "wiggly." These wiggles, somewhat exaggerated by the vagaries of the machine, are a manifestation of *statistical fluctuations*. And these small capricious fluctuations are all that stands between us and the seemingly inevitable heat death.

Enough has been said to convince one that the vague nonmathematical notions of chance, randomness, and independence can be made precise and become incorporated into a formal mathematical system. In spite of this, an impression may linger, as indeed it keeps lingering among nonmathematicians, that probability theory, because of its concern with chance phenomena, is somehow set apart from other mathematical disciplines. To counteract this impression, we conclude with two instructive examples.

The first concerns Buffon's needle but in a different setting, for, instead of throwing the needle "at random," we shall now endow it with a perfectly predictable motion in which the center of the needle has a constant velocity component v in the direction perpendicular to the floor cracks,

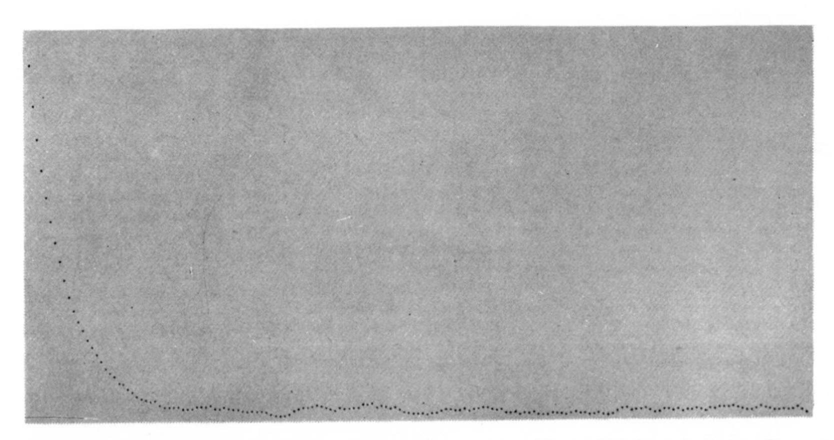

FIGURE 12. *Played on a computer, an Ehrenfest game with 16,384 hypothetical balls and 200,000 drawings took just two minutes. Starting with all the balls in container* A, *the number of balls in* A *was recorded with a dot every 1,000 drawings. It declined exponentially until equilibrium was reached with 8,192 balls (half of them) in each container. After that fluctuations were not great.*

while at the same time the needle rotates about its center with constant angular velocity of ω revolutions per second. If we now observe the motion for a long time and calculate the fraction of the time during which the needle is intersecting some floor crack, we shall find that it will be very close to Buffon's answer $2l/\pi d$, provided that the number $v/d\omega$ is irrational.

In this formulation, nothing is "random" and nothing is left to "chance." That we nevertheless get the same answer as before is not an accident but a consequence of the fact that the motion of the needle happens to provide us with a realization of assumptions that were needed to derive Buffon's answer. In particular, the irrationality of the ratio $v/d\omega$ is responsible for the independence of x and θ.

The second example is perhaps even more decisive. Take the consecutive integers $1, 2, 3, 4, 5, \ldots, n$ (you may think of n as being 10,000 to be more specific), and corresponding to each integer m in this range, consider the number $\nu(m)$ of its different prime factors.

Thus $\nu(1) = 0$, $\nu(2) = \nu(3) = \nu(5) = \cdots = 1$, $\nu(4) = \nu(2^2) = 1$, $\nu(6) = \nu(2.3) = 2$, $\nu(60) = \nu(2^2.3.5) = 3$, etc.

I can now construct a histogram of the number of prime divisors. If I center it at log (log n) (logarithms being the natural logarithm of John Napier) and adjust suitably the size of the base interval, I can again with excellent accuracy (which improves as n gets larger and larger) approximate the histogram by the normal curve. The statistics of the number of prime factors is thus indistinguishable from the statistics of sizes of peas or the statistics of displacements in Brownian motion. And yet again there is no "chance" and no "randomness."

It is, of course, impossible in the available space to give an adequate account of probability theory. Whole areas of importance and current interest (like diffusion theory, information theory, or the theory of queues, to name but a few) had to be left out. We could not touch upon the crucial question of experimental verifiability of predictions based on probability theory. And we have had to leave out all mention of mathematical statistics in spite of its close ties with probability theory and its universally acclaimed place in scientific methodology.

But the brief glimpses of the few problems and developments we were able to include were perhaps sufficient to convince you of the unique place that probability theory occupies not only in mathematics but in science as a whole.

On the one hand, it is a thriving branch of pure mathematics; on the other, it is an indispensable tool of an engineer. On the one hand, it helps us learn something new about prime factors of integers; on the other, it tempers the pessimism of the Second Law with the hope born out of fluctuations.

Laplace must have foreseen the immense potentialities of the theory he helped so much to create when he wrote, in his *Théorie Analytique des Probabilités*, that "the most important questions of life are, for the most part, really only problems of probability."

BIBLIOGRAPHY

William Allen Whitworth, *Choice and Chance: With One Thousand Exercises*, Hafner, New York, 1951.

I. Todhunter, *A History of the Mathematical Theory of Probability from the Time of Pascal to That of Laplace*, Chelsea, New York, 1949.

William Feller, *An Introduction to Probability Theory and Its Applications*, Volumes I and II, Wiley, New York, 1957.

Warren Weaver, *Lady Luck*, Doubleday, Garden City, N.Y., 1963.

Richard von Mises, *Probability, Statistics and Truth*, Macmillan, New York, 1957.

Samuel Goldberg, *Probability: An Introduction*, Prentice-Hall, Englewood Cliffs, N.J., 1963.

BIOGRAPHICAL NOTE

Mark Kac is professor of mathematics at Rockefeller University (formerly the Rockefeller Institute) in New York City. Born in Poland, he came to the United States in 1938 and, after a year at The Johns Hopkins University, joined the faculty at Cornell University, where he remained for more than twenty years. His interest in probability started when one of his professors pointed out that the properties of some well-defined mathematical entities like functions are analogous to ideas encountered in the world of chance. Later on, his interest in probability was reinforced by his preoccupation with problems of statistical mechanics.

How large is infinity? A good many ideal mathematical objects are infinite in number, for example, the positive whole numbers or the points on a line segment. Until fairly recent times mathematicians made no distinctions among them. Infinity was simply infinity. Now, however, they know that in the peculiar "arithmetic" of infinite sets the number of points on a line represents a larger infinity than the number of positive integers. But is there an infinite number intermediate between these two? A widely accepted assumption, called the "continuum hypothesis," says there is no such infinite number. But this assumption, far from being proved, remains one of the most ticklish questions in mathematical logic. Attempts either to prove or disprove the continuum hypothesis lead mathematicians deep into the roots of mathematical thought.

The Continuum Hypothesis

Raymond M. Smullyan

The famous Continuum Hypothesis — about which there is a recent revival of interest — arose with Cantor's theory of sets. A set is to be thought of as any collection of objects whatsoever. Set theory is very prominent today because it appears that all mathematics can be developed starting with the notions of "set" and an object's being a member of a set. (The remaining undefined notions of mathematics belong to pure logic.) Yet today we are still totally in the dark concerning the truth or falsity of one of the absolutely fundamental questions about sets, namely the Continuum Hypothesis. Before we can even state what this hypothesis is, we must backtrack a bit to Cantor's theory about infinite sets.

What does it mean to say that two sets contain the same number of elements or that one set has a greater or lesser number of elements than another? For *finite* sets, the answer is easy. Let us illustrate this with the following example.

Suppose we look into the window of a theater, and we see that every seat is occupied by just one person and also that no one is standing. Then, without having to count either the number of seats or the number of people, we know that the numbers are the same, because the set of seats can be put into a *one-to-one* correspondence with the set of people. On the other hand, if we see all seats occupied and also some people standing, we can say that there are more people than chairs, since the set of chairs is in a 1–1 correspondence with a part of but not the whole of the set of people. In mathematical language, the set of chairs is in a 1–1 correspondence with a *proper subset* of the set of people. We might also have the situation that everyone is seated but some of the seats are vacant. Then the set of people can be put into a 1–1 correspondence with a proper subset of the set of chairs, in which case we say that there are more chairs than people.

Thus, for *finite* sets S_1, S_2, the definition of what it means for them to have the same number of elements or for one of them to have a lesser number of elements than the other is simple: We say S_1 has the same number of elements as S_2 if S_1 can be put into a 1–1 correspondence with S_2 and that S_1 has a lesser number of elements than S_2 if S_1 can be put into a 1–1 correspondence with a proper subset of S_2. (For *infinite* sets, however, this definition of "lesser" is valueless, as we shall soon see.) The important fact is that for finite sets S_1, S_2, if there is one way of putting S_1 into a 1–1 correspondence with S_2, there is no way of putting S_1 into a 1–1 correspondence with a proper subset of S_2.

To return to our theater example, suppose the original situation prevails; that is, everybody is seated and there are no empty seats. Now suppose we ask all the people to stand up, wander about, and choose new seats at random. After they do this, it is simply not possible that everyone is seated and some seats are vacant or that all seats are taken but someone is standing. In other words, the particular seating arrangement has no bearing on whether there are too few seats, too many seats, or exactly the right number of seats.

Matching Infinities

With *infinite* sets, however, the situation is very different. Imagine (if you can) an *infinite* number of seats arranged in a row and numbered 1, 2, 3, . . . , n, . . . , going to the right, with each seat taken and nobody standing. Then the set S_1 of seats is in a 1–1 correspondence with the set S_2 of people. Now suppose we ask everybody to move over one seat to the right. Then we have the situation where everybody has a seat, yet there is one seat vacant (namely the first seat), and so we have put a proper subset of S_1 into a 1–1 correspondence with all of S_2.

Alternatively, instead of asking everyone to move over one seat to the right, we could ask the person in the first seat to stand up and then ask everyone to move one seat to the left. Now we have the situation where all seats are taken, yet there is someone standing; therefore, we have put S_1 into a 1–1 correspondence with a proper subset of S_2. So we see that, for infinite sets, S_1, S_2, one particular 1–1 correspondence might be from all of S_1 to all of S_2, another might be from all of S_1 to a part but not the whole of S_2, and yet another might be from a proper part (that is, a part but not the whole) of S_1 to all of S_2. Under these circumstances, should we say that S_1 has the same number as or fewer or more elements than S_2?

The answer is this: As with finite sets, we say that S_1 has the same number of elements as S_2 if S_1 can be put into a 1–1 correspondence with S_2. The delicate point now is the definitions of "lesser" and "greater." We say that S_1 is less (or has fewer elements) than S_2 if there does exist at least one 1–1 correspondence from S_1 to a proper part of S_2 and, furthermore, there does not exist any 1–1 correspondence from S_1 to all of S_2. According to this definition, we see that in our infinite theater example, the number of seats is the same as the number of people, since there is at least one way of seating all the people so that all seats are taken.

If two sets S_1 and S_2 can be put into a 1–1 correspondence, we say that S_1, S_2 have the same cardinality; if S_1 has fewer elements than S_2, we say that S_1 has a lower cardinality than S_2 or that S_2 has a higher cardinality than S_1.

We might now consider the following question: Suppose S_1 can be put into a 1–1 correspondence with a part of S_2 and S_2 can be put into a 1–1 correspondence with a part of S_1. Does it necessarily follow that all of S_1 can be put into a 1–1 correspondence with all of S_2? The answer is "yes." This result is known as the Bernstein–Schröder theorem. (The reader can consult [1, Chapter XII] for a proof.) Another natural question to consider is whether every pair, S_1, S_2, of sets is necessarily *comparable*, that is, whether either S_1 can be put into a 1–1 correspondence with a part of S_2 or S_2 can be put into a 1–1 correspondence with part of S_1. This is equivalent to the question of whether, for any two sets S_1, S_2, either S_1 has lower, higher, or the same cardinality as S_2. The answer is again "yes," but the proof depends on the acceptance of a mathematical principle with a rather special status; this principle is known as the *Axiom of Choice*. This axiom (in one form) says that, if C is a nonempty collection of nonempty sets and if no two distinct members of C have a common element, there is a set S (a so-called "choice" set) that contains exactly one element from each member of C. We will later have more to say about the Axiom of Choice.

Counting to Infinity

Perhaps the first basic question about infinite sets is whether or not there exist infinite sets S_1, S_2 with different cardinalities. Stated otherwise, could it not be true that on the one hand we have finite sets and on the other we have infinite sets and that every finite set has fewer elements than every infinite set but that any two infinite sets have the same number of elements? A set is called *denumerable* if it can be put into a 1–1 correspondence with the set of positive natural (that is, whole) numbers $1, 2, 3, \ldots, n, \ldots$. The question then can be equivalently posed: Is every infinite set denumerable or are there *nondenumerable* infinite sets?

Cantor devoted much time and thought to this question. He gave many interesting and important examples of sets which at first sight appeared to be nondenumerable but are really denumerable. We will illustrate some of Cantor's constructions by the following game.

Imagine that you are immortal, and I say to you that I am thinking of a particular positive integer 1, 2, 3, . . . , n, ... Every day you are allowed one guess as to the number; if and when you guess it, you get a prize. Does there exist a method whereby you can surely win on some day or other? The strategy in this case is completely obvious: On the first day you guess "1," on the second you guess "2," and so on. Sooner or later you are bound to win. Now I make the game a wee bit harder. I announce that I am thinking either of some positive integer 1, 2, 3, . . . or some negative integer -1, -2, -3, ... Again you are allowed only one guess a day. How can you be sure and win this time? If you guess "1" on the first day, "2" on the second, and so on, and if I happen to be thinking of a negative integer, you will never get around to naming it. Or if you should guess in the order -1, -2, . . . , and I happen to be thinking of a positive integer, again you will never win. But a moment's thought will reveal that, if you guess in the order 1, -1, 2, -2, 3, -3, . . . , you are bound to win. You have thus put the set S consisting of all positive and all negative integers into a 1–1 correspondence with the set of positive integers alone — or, as we say, you have *enumerated* the set S. Thus you have shown that the set S is denumerable, even though at first sight it may appear "twice as big" as the set of positive integers alone. Now I make the game considerably harder. This time I am thinking of a positive *rational* number (called also a *fraction*), that is, a number of the form m/n, where

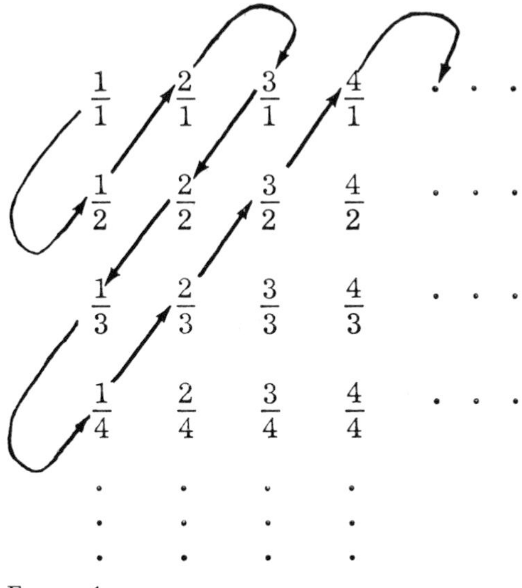

FIGURE 1

m, n are positive whole numbers. What is your strategy now? The situation might seem really hopeless since there are infinitely many numerators to choose from and, with each choice of a numerator, there are infinitely many possible choices of a denominator. Yet the set R of positive rationals can be enumerated by the plan shown in Figure 1.

Nondenumerable Sets

These examples, and many others, led Cantor to conjecture that every infinite set is denumerable. Indeed, Cantor spent many years trying to prove this conjecture. Then Cantor made the basic discovery that the conjecture was false. More specifically, Cantor showed the following. For any set S, the *power set* of S — written $P(S)$ — means the set of all subsets of S. We shall let N stand for the set $1, 2, 3, \ldots, n, \ldots$ of positive whole numbers. Cantor showed that the set $P(N)$ of all sets of integers is nondenumerable (that is, cannot be put into a 1–1 correspondence with N). More generally, Cantor showed that, for any set S, $P(S)$ has higher cardinality than S. (For finite sets, this was, of course, well known since, if S has n elements, $P(S)$ has 2^n elements). For infinite sets S, Cantor's famous argument is both remarkably ingenious and remarkably simple, and it runs as follows.

Consider any 1–1 correspondence that assigns to each element a of S an element, which we will call K_a, of $P(S)$. In other words, each element a of S corresponds to a subset K_a of S. We wish to show that the correspondence is not onto the whole of $P(S)$ but only onto a proper subset of $P(S)$. In other words, we must exhibit an element K of $P(S)$ such that no element a of S corresponds to K. Stated otherwise, we seek a K such that, for every a in S, K_a is distinct from K. Now, for any element a of S, it might happen that a is a member of the set, K_a, to which it corresponds, or again it might not happen. In any case, either a belongs to K_a or it doesn't. Well, we take for K the set consisting of just those elements a such that a does not belong to K_a. Then the two sets K, K_a are distinct from each other, because one of them contains the element a and the other does not. Thus, for every a in S, $K \neq K_a$, so K has the property we seek. This proves that S cannot be put into any 1–1 correspondence with all of $P(S)$. Yet S can certainly be put into a 1–1 correspondence with part of $P(S)$ — just assign to each element a of S the set $\{a\}$ whose only element is a. Thus, $P(S)$ does indeed have higher cardinality than S, and Cantor's theorem is proved.

In particular, the set $P(N)$ of all subsets of N is nondenumerable (has higher cardinality than N). It can be shown that $P(N)$ can be put into a 1–1 correspondence with the set of all points on a given straight line; for this reason, $P(N)$ is said to have the cardinality of the *continuum*. Now, the Continuum Problem is this: Does there exist a set of higher cardinality than N but of lower cardinality than $P(N)$? The *continuum hypothesis* asserts that there is no such intermediate set. And the so-called *generalized continuum hypothesis* asserts that, for every set S, there never can be a set of higher cardinality than S but of lower cardinality than $P(S)$.

Of Truth and Consistency

Today many principles are known to be consequences of the generalized continuum hypothesis, and many principles are known to be equivalent to this hypothesis. But we have absolutely no idea whether the continuum hypothesis itself is true or false. Another question, not to be confused with the truth or falsity of the continuum hypothesis, is whether it can be formally proved or disproved from the present-day axioms of set theory. This question is now completely settled.

We use the term "present-day axioms of set theory" to mean the system ZFS (Zermelo-Fraenkel-Skolem), which is the system currently in most widespread use. In 1938 Kürt Gödel proved the famous result that the generalized continuum hypothesis is *formally consistent* with the axioms of ZFS (assuming, of course, that ZFS is itself consistent, *which we shall assume*). And in 1963 Paul Cohen [2], settled the matter in the other direction; that is, he showed that the negation of the generalized continuum hypothesis — in fact, even the negation of the special continuum hypothesis — is consistent in ZFS. Thus, the continuum hypothesis is *independent* of the axioms of ZFS; it can be neither proved nor disproved in ZFS. So the axioms of ZFS are not strong enough to settle the continuum problem.

Another important independence result concerns the axiom of choice. Gödel has shown also that the axiom of choice is not disprovable in ZFS, and Cohen has shown that it is not provable in ZFS. Hence, the axiom of choice is also independent of the axioms of ZFS. The most remarkable result of Cohen is that, even if we add the axiom of choice to the other axioms of set theory (that is, to ZFS), it is still not possible to prove the continuum hypothesis.

There has been a remarkable diversity of opinions concerning the significance of the independence results. First of all, we must realize that these results have been proved specifically for the particular axiom system ZFS, though the arguments hold for many related systems. There are, however, very different formal axiom systems of set theory in which Gödel's and Cohen's arguments do not go through. But it is highly questionable whether these other systems really describe the notion of "set" as the working mathematician uses it. (In Quine's system [3], for example, the axiom of choice is provably false.) For one who regards all these alternative systems to be on an equal footing, and this includes many so-called formalists, the independence results may well seem insignificant; they could be construed as merely saying that the continuum hypothesis comes out positive in one system and negative in some other, equally good system. On the other hand, there are those who concede that ZFS is the more "natural" set theory but who cannot understand what could be meant by mathematical truth other than provability in ZFS. They would, therefore, construe the independence results as saying that the continuum hypothesis is neither true nor false. A slightly modified viewpoint is that the only propositions we can ever know to be true are those provable in ZFS and, hence, that we can never know whether the continuum hypothesis is true or false.

The so-called mathematical realist, or Platonist (and this seems to include a large number of working mathematicians), looks upon the matter very differently. We can describe the realist viewpoint as follows. There is a well-defined mathematical reality of sets, and in this reality the continuum hypothesis is definitely true or false. The axioms of ZFS give a true but incomplete description of this reality. The independence results cast no light on the truth or falsity of the continuum hypothesis; nor do they in any way indicate that it is neither true nor false. Rather, they highlight the inadequacy of our present-day axiom system ZFS. But it is perfectly possible that new principles of set theory may one day be found which, though not derivable from the present axioms, are nevertheless self-evident (as the axiom of choice is to most mathematicians) and which might settle the continuum hypothesis one way or the other. Indeed, Gödel — despite his own proof of the formal consistency of the continuum hypothesis — has conjectured that, when such a principle is found, the continuum hypothesis will then be seen to be false.

In the interests of the present exposition, we shall adopt a realist viewpoint.

Constructible Sets

Gödel proved the consistency of the axiom of choice and the generalized continuum hypothesis by defining a certain type of set called a *constructible* set. Briefly, a set is constructible if it belongs to every model of ZFS of the same size as the universe of all sets. (In [4] can be found a more precise account.) The first important fact about constructible sets is that if we reinterpret the word "set" to mean "constructible set," all the axioms of ZFS remain true under this new interpretation. We paraphrase this by saying that the constructible sets form a model for ZFS; the symbol "L" is used to denote this model. Now, under this new interpretation of "set," various notions of set theory, all defined ultimately in terms of set, are liable to change in meaning; a notion (or definition) is called *absolute* if it does not undergo such a change in meaning. An enormous number of notions in set theory turn out to be absolute. In particular, the very notion of constructibility is absolute. This means that the constructible sets are not only constructible but are even constructible in the new sense — in other words, the very statement "all sets are constructible" (which is called the *axiom of constructibility*) is true in the model L. So we have a model L in which the axioms of ZFS as well as the axiom of constructibility hold. This shows that the axiom of constructibility is consistent with the other axioms of set theory since there is a model for both — again, of course, assuming all the while that ZFS itself is consistent.

The second important fact about constructible sets is that the axiom of constructibility implies the generalized continuum hypothesis as well as the axiom of choice. Since the axiom of constructibility is true in the model L, so is the continuum hypothesis. In other words, if we reinterpret "set" to mean "constructible set," the generalized continuum hypothosis becomes true. Thus Gödel shows that the generalized continuum hypothesis is consistent relative to ZFS.

Cohen, of course, shows the consistency of the negation of the continuum hypothesis by giving a model of ZFS in which the continuum hypothesis is false. This model is based on a very significant idea (introduced by Cohen) known as *forcing*, which comes extremely close to certain basic semantic notions of *intuitionistic* logic. (The reader should consult an account of Cohen's work, *Set Theory and the Continuum Hypothesis* [2].) Cohen also gave a model for ZFS in which the axiom of constructibility is false. Thus, the axiom of constructibility, like the axiom of choice and the generalized continuum hypothesis, is independent of the axioms of ZFS. The truth (or falsity) of the axiom of constructibility is still as unknown as that of the continuum hypothesis.

Although the truth or falsity of the axiom of constructibility is not known, rather recently new light has been thrown on the matter by the consideration of certain very large sets known as *measurable cardinals*. We let MC be the hypothesis that measurable cardinals exist. The first important relation between the hypothesis MC and the axiom of constructibility was discovered by Dana Scott. He showed the extremely surprising result that MC implies the existence of nonconstructible sets. Subsequently, Rowbottom (doctoral dissertation, University of Wisconsin) sharpened the result and showed that the existence of a measurable cardinal implies the existence of a nonconstructible set of integers; better still, he showed that MC implies that there are only denumerably many constructible sets of integers.

Since MC succeeded in throwing light on the axiom of constructibility, there naturally arose hopes that MC might throw light on the continuum hypothesis. This hope has very recently been dispelled. In late 1964 Azriel Levy in Jerusalem and Robert Solovay in Princeton, independently and nearly simultaneously, showed (using Cohen-like techniques) that, even if we add MC and the axiom of choice to the axioms of ZFS and if the resulting system is consistent, then it is still not possible to prove or disprove the special continuum hypothesis. It is not known whether this holds for the general continuum hypothesis.

The consistency of MC itself (in ZFS) remains an unsolved problem.

BIBLIOGRAPHY

1. R. L. Wilder, *Introduction to the Foundations of Mathematics*, Wiley, New York, 1965.
2. Paul Cohen, *Set Theory and the Continuum Hypothesis*, W. A. Benjamin, New York, 1966.
3. W. V. Quine, *Mathematical Logic*, W. W. Norton, New York, 1940.
4. Raymond M. Smullyan, "The Continuum Problem," *The Encyclopedia of Philosophy*, Macmillan and the Free Press, New York, 1967.

BIOGRAPHICAL NOTE

Born in New York, Raymond Smullyan became interested in the foundations of mathematics at a very early age. His graduate training was taken at Princeton University, and then he taught at Belfer Graduate School of Science, Yeshiva University. At present he is a professor of mathematics at the City University of New York. Theory of formal systems, recursive functions, and proof theory are the main areas in which he has worked. Among Professor Smullyan's nonmathematical interests, music and magic are prominent: He has developed professional-level skills as a pianist and worked his way through graduate school as a professional magician.

For all its complexity, the electronic computer is an instrument for simplifying problems that overwhelm the intellect. In this respect, it has the same broad goal as traditional mathematics, but the approach is diametrically different. While the mathematician searches for a gem of clarity amid a welter of information, a computer attacks problems like a bulldozer. Faster machines with more capacious memories and much more autonomous programs are due soon to increase greatly the raw power of the computer. The author intimates that what has occurred so far in computer science may be but the faintest indication of what is to come. Much research is devoted to what might be called basic problems in thinking. Out of this may come eventually a new order of simplifying complexity. While speculating on the possibility of such computer-controlled organisms as automatically motivated factories, the author foresees the evolution of a new form of lifelike machines. If machines are ever endowed with human intelligence, it will be relatively easy to create superintelligences.

Prospects of Computer Science

J. T. Schwartz

Computer science, a new addition to the fraternity of sciences, confronts its older brothers, mathematics and engineering, with an adolescent brashness born of rapid, confident growth and perhaps also of enthusiastic inexperience. At this stage, computer science consists less of established principles than of nascent abilities. It will, therefore, be the aim of this essay to sketch the major new possibilities and goals implicit in the daily flux of technique.

The power of computers is growing impressively. A few years ago, they carried out thousands of instructions per second. Today's models are capable of tens and hundreds of millions of instructions per second. This rapid advance is expanding the unique ability of the computer: its power to deal successfully with the irreducibly complex — that is, situations and processes that can be specified adequately only by very large amounts of information.

Without the high-precision and high-speed tool that is the computer, a process that is irreducibly complex in the above sense is unusable (and hardly conceivable) for two fundamental reasons. First, the accumulation and verification of great volumes of highly accurate interdependent information is not practical. For instance, because of inevitable human error in their preparation, completely accurate mathematical tables did not exist before computers became available. Second, since complex processes demand long sequences of elementary steps, no really complex process can actually be carried through without computers.

These two obstacles present themselves at so fundamental a level in any intellectual investigation that until the development of computers the possibility of dealing successfully with the complex itself was never really envisaged. Perhaps the most successful substitute for such a possibility, as well as the nearest approach to it, came in mathematics. Mathematics provided simplification in a different way: by furnishing intellectual techniques of conceptualizing abstraction and developing symbolic and nonsymbolic languages to facilitate the manipulation and application of concepts. Utilizing these techniques, mathematics has developed an impressive ability not really to deal directly with the complex but, rather, to hunt out ways to reduce complex content to principles that are briefly statable and transparent. To find the simple in the complex, the finite in the infinite — that is not a bad description of the aim and essence of mathematics.

Since the linguistic-conceptual procedure of mathematics is the most powerful method for helping the unaided intellect deal with precisely defined situations, the concepts and language of mathematics have been profitably adopted by many sciences. Beyond the individual strong points occupied by mathematics and marked out by its famous theorems and definitions lies a jungle of combinatorial particularities into which mathematics has always been unable to advance. Computer science is beginning to blaze trails into this zone of complexity. To this end, however, methods are required which, while they include some of the classical devices of mathematics, go beyond these in certain characteristic respects.

In this quest for simplification, mathematics stands to computer science as diamond mining to coal mining. The former is a search for gems. Although it may involve the preliminary handling of masses of raw material, it culminates in an exquisite item free of dross. The latter is permanently involved with bulldozing large masses of ore — extremely useful bulk material. It is necessarily a social rather than an individual effort. Mathematics can always fix its attention on succinct concepts and theorems. Computer science can expect, even after equally determined efforts toward simplification, only to build sprawling procedures, which require painstaking and extensive descriptive mapping if they are to be preserved from dusty chaos.

In its initial efforts, information science has of course, followed the easiest path, emphasizing those processes leading to results that are perhaps too complex for direct analytic treatment but can still be attained by use of relatively simple algorithms iteratively applied. These processes are complex only in the limited sense that carrying them out requires long sequences of simple steps. Most of the calculations, data analyses, simulations, and so forth, which are the common coin of present-day computer work, are large-scale repetitions of this kind. Elaborate programs of the types of language translators, linguistic analyzers, and theorem provers lie more definitely within the borders of the newly opened region of complexity into which computer science will progress as its equipment and technique grow stronger.

Technological Past and Perspectives

The modern computer is, of course, a *stored program* computer. That is, it includes a mass of recorded information, or *memory*, stored temporarily on any medium the engineer finds attractive, for example, magnetic tape, wired arrays of small circular magnets, punched cards, punched paper tape, celluloid strips. It also has one or more *processing* units, whereby segments of stored information are selected and caused to interact. The new information produced by this interaction is then again stored in memory, a procedure that allows successive steps of the same form to be taken.

Three separate segments of information are generally involved in each elementary interaction. Two of these are usually treated as coded representations of numbers to be combined by arithmetic or logical operations. The third is treated as a coded indication of the memory location from which these operands are to be drawn and of the operation to be performed: addition, multiplication, comparison, and so on. Such an elementary combination, or *instruction execution*, produces not only a direct result but also an indication of the place in memory where the result is to be recorded and an indication of the place where the next instruction is to be found. The power of a computer is roughly proportional to the rate at which it executes these elementary combinations.

A number of major recent technological developments have combined to advance computer power rapidly in the last decade.

1. The development of commercially available ultra-high-speed electronic circuits has made it practical to execute the elementary instructions in tenths or even hundredths of millionths of a second. The high reliability of transistors has made it possible to keep enormous assemblies of parts functioning simultaneously for extended periods. The development of "integrated" microelectronics, built up by the diffusion of electric catalysts into silicon or germanium chips, has reduced the size of computers, permitting signals to pass between their extremities in short periods. Computers now in design will be able to coordinate and to carry out a multiplication, a pair of additions, and a logical operation in the time it takes light to travel 15 feet, sustaining sequential instruction rates of 200 million instructions per second.

2. The development of manufacturing technique has reduced the cost of storing information in rapidly accessible form. Increasing sophistication in the over-all configuration of computer systems has made possible the

effective and economic combination of available information storage media. The large computer system of the near future will have a hierarchy of rapidly accessible memories. A few hundred digits will be stored directly in high-speed electronic circuits, from which they may be read in some hundredths of a millionth of a second. Half a million to a few million digits in bulk electronic storage will be readable in well under $\frac{1}{10}$ of a millionth of a second. And as many as 10–200 million digits in lower-grade magnetic storage will be accessible in $\frac{1}{4}$ of a millionth of a second or less. Such systems will also incorporate a variety of newly perfected bulk information storage devices, ranging through juke-box-like magnetic disk cabinets with fixed or movable recording arms, storing some hundreds of millions of digits accessible in tenths or hundreths of a second, plastic-strip information files storing billions of digits, and elaborate photographic microdot cabinets, in which trillions of digits can be stored for access in a few seconds. Such systems will constitute very impressive information-processing factories, in which an information base comparable in size to the Library of Congress can be processed at the rate of hundreds of millions of instructions per second.

3. Increased sophistication in computer design has yielded substantially increased performance from given quantities of electronic circuitry. By artfully overlapping the execution of as many instructions as possible, a large part of the circuitry comprising a computer may be kept steadily in action. Ingenious techniques have been devised for the dynamic movement of data between the various storage media available to a computer to permit more effective use of the memory. Special arrangements permitting the computer to hold many tasks in immediate readiness, so that calculation may proceed on one while data are being fetched for others, also contribute to the productivity of a given computer complex.

4. The continued rapid improvement of manufacturing technique and electronic reliability should soon make possible the construction of considerably larger circuit assemblages. These larger computers could, by allowing more parts of a logical process to proceed in parallel, attain very greatly increased instruction rates — up to several thousand million instructions per second. Improved manufacturing techniques should also make possible considerable increases in the capacities of fast computer memories and should eventually make possible the construction of memories that are themselves capable of some degree of logical activity.

5. The availability of faster computers supplied with considerably increased banks of memory aids significantly in simplifying programming. In programming an algorithm to run on a small slow computer, considerable effort is needed to compress the code and increase its efficiency. "Hand coding," reflecting this necessity, is more complex than mechanically generated codes would have to be to accomplish the same purpose. In some cases, this additional complication is extreme, and an effort quite disproportionate to the simplicity of a basic algorithm goes into the

design of specialized layouts of data structure and special schemes for reducing the number of passes over a tape which a calculation will require. Improved machines should progressively free programming from the burden of such cares. As programmers adjust their habits to improved equipment, their ability to cope with inherently complex situations should improve.

6. Not only the physical computer but also the programming art has advanced during the past decade. An early but very basic concept is that of the *subroutine linkage*, whereby a generally useful subprocess — for example, taking a square root — could be separately programmed and routinely linked into a more comprehensive process. This technique makes it possible for an elaborate algorithm to be analyzed into simpler parts, which may then be programmed individually. Conversely, complex processes may rapidly be built up by the hierarchical combination of much more elementary logical parts. The utility of such a method is enhanced by its embodiment in a *programming source language*. Such a language is an assemblage of standardized elementary processes together with a grammar permitting the combination of these into composites. By standardizing programming technique, by providing for the automatic treatment of a host of detailed questions of data layout and access, by establishing succinct, mnemonic, and easily combined representations for frequently used processes, and by automatically checking programs for internal consistency and grammatical legality, such source languages make programming very much more effective than it would otherwise be.

The possibility of communicating with a computer through a programming language comes, of course, from the fact that the instructions a computer executes have the same internal format as the numbers with which it calculates. Thus, given the algorithm that describes the manner in which a particular programming language is to be expanded into a detailed set of internal instructions, a computer can calculate its own instructions from the indications present in a source language program. Moreover, since the phrases of a source language may be coded for internal representation within a computer, computers can also manipulate the statements of a programming language. Furnished with suitable algorithms, the computer is then able to accept definitions extending the grammar and vocabulary of a programming language and to retranslate extended statements into the basic statements they represent. Use of this technique permits an initial source language to be developed by successive extension for effective application to a given problem field. Searching for patterned repetitions in a given algorithm and embodying these patterns suitably in definitions may make it possible to express the algorithm efficiently and succinctly in a suitably extended programming language.

Another advantage of using programming languages for the communication of instructions to computers is that these languages are relatively machine-independent. Computers differing very greatly in their internal structure and repertoire of instructions can all be furnished with the same programs, each computer translating the source language into its own internal language. In particular, transition from a given computer to a later improved computer need require relatively little reworking of programs.

7. Experience has shown that various data structures (trees, lists, and so on) more complex than mere linear or multidimensional data arrays can be used effectively in the programming of elaborate algorithms. Various languages embodying useful general procedures for dealing with such structures have been constructed. Interesting methods have been developed for representing both complex data structures and algorithms for processing them with lists of a single universal form. Such methods preserve the complete flexibility of the underlying computer but allow its use at a considerably more organized level than that of internal machine language.

8. The continued development of source languages has made it possible to describe them to a computer largely in terms of the language itself. While the most basic elementary terms of any such description must be stated in a manner reflecting the specific instruction repertoire of a given computer, this "machine-dependent" portion of the description can be confined to a rather small part of the total information specifying the language. Thus, once the language is made available on a given computer, it can be carried over to a second computer by a relatively few redefinitions of basic terms, from which the first computer can proceed to regenerate the whole translation program in a form directly suitable for use on the second. If this "bootstrapping" technique, presently in a state of development and experimental use, could be suitably strengthened by the inclusion of program optimization algorithms, it might become possible for the translator to pass from a second computer to a third, and so on, and then back to the first without substantial loss of efficiency. Since each translation program carries along with it the body of programs written in the source language it translates, it would then become easier to assure continuity of development of the programming art.

The Computer as an Artificial Intelligence

The computer is not an adding machine but a universal information-processing engine. Its universality follows from the following basic observation of Turing: Any computer can be used to simulate any other. Thus, what one computer can do another can do also. Any computer is but a particular instance of the abstractly unique universal information processor. Computers differ not in their abstract capacities but merely in practical ways: in the size of the data base the computer can reach in a given time and in the speed with which a given calculation can be accomplished. Turing's observation applies, by virtue of its simplicity and generality, to any information-processing device or structure subject to the causal laws of physics and, thus, to any information-processing device which is wholly natural rather than partly supernatural. It should in principle apply, therefore, to animal brains in general and to the human brain in particular.

From these reflections, one may derive a fascinating provisional goal for computer research: to duplicate all the capabilities of the human intelligence.

A crude inventory of the principal abilities that together constitute living intelligence will indicate the gross steps necessary in creating artificial intelligence.

1. The human brain contains some 10 billion nerve cells, each capable of firing or failing to fire some hundreds of times per second. We may thus estimate that the information-producing capability of the brain is equivalent to the processing of approximately 1–10 trillion elementary bits of information per second. A computer of the largest current type contains some 500,000 elementary logical circuits, each of which may be used approximately 4 million times per second. Thus it processes approximately 2 trillion bits per second and, hence, is probably comparable to the brain in crude information-processing capability. Of course, the fact that individual neurons can perform rather more complex combinations of incoming signals than are performed by elementary computer circuits means that the foregoing estimates are uncertain, possibly by a factor of 100. Even if this is the case, however, computers available in the next few years should equal or exceed the crude information-processing capacity of the brain.

Access rates to memory are, however, much larger in the brain than in computers. Even assuming that the average neuron stores only 1 elementary bit of information, the brain has access to 10 billion bits, or 3 billion digits, of information at rates of access that do not impede its normal rate of function. Even a large computer of present design will not have access to more than 3 million digits of information at rates relatively as rapid. While much larger masses of information are, in fact, available to the computer, they are available only after substantial access delay.

Thus, while the ability of a large computer to process data probably equals that of the brain, its access to data is probably only 0.1 percent as satisfactory as that of the brain. These considerations reveal why computers perform excellently in iterative calculation with small amounts of data and in processing larger amounts of information used in simple sequences but are quite poor compared with the living brain in carrying out highly sophisticated processes in which large amounts of data must be flexibly and unpredictably combined. A considerable development of manufacturing technique will be required before it becomes feasible to furnish a computer with a rapid access memory more nearly comparable to that of the living brain.

A considerable development of the programming art also will be necessary before impressive artificial intelligences can be constructed. While it is true that much of the information used during the functioning of the adult brain is learned during earlier experience, it may also be surmised that a very large amount of genetic information contributes to the capacity to learn rapidly from fragmentary evidences. If, as a crude guess, we assume this inherent information comprises 10^{10} elementary bits, one for every neuron in the brain,[1] we may estimate the programming labor necessary to duplicate this information. It would be an excellent programmer who by present techniques is able to produce 300,000 correct instructions (10 million elementary bits of information) per year. Thus, one might estimate that to program an artificial intelligence, even assuming the general strategy to be clearer than it really is at present, is the labor of 1,000 man-years and possibly a hundred times more.

2. In the retina of the eye, the optic nerve, and the visual cortex of the brain the rapid and very extensive processing and reduction of the total pattern of light and dark sensed by the individual receptor cells of the eye take place. By a process currently understood only in bare outline, a vast mass of incoming visual data is reduced to a much smaller set of gestalt fragments or descriptor keys, which can be compared against an available stock of visual memories for recognition as a familiar object. A similar data reduction is performed in decoding auditory signals and is probably associated with the other senses as well.

The development of similarly sophisticated input programs for computers has been under way for some years and has attained modest success. It has become possible for computers equipped with optical scanners to read printed text with reasonable rapidity. A considerable effort in automatic spark-chamber and bubble-chamber photograph analysis has succeeded in developing programs capable of recognizing particle tracks and characteristic nodes in such photographs and of going on to analyze the geometry of perceived collisions. The evident economic importance of this visual pattern-recognition problem and the availability of faster computers better supplied with memory for experiments in pattern recognition assure continued progress in this area.

3. The brain coordinates the activity of elaborate assemblages of muscles in the apparatus of speech and in the body at large. This effector control is combined with sophisticated analysis of each momentary pattern of autosensation, permitting the stable and effective performance of tasks in a varying environment. The absence until two decades ago of mechanical brains of any kind other than crude mechanical clockworks and rudimentary electric feedback circuits has unfortunately caused

[1] Note, for comparison, that a person with perfect recall, all of whose information was derived from reading at the rate of 200 pages per hour, 20 hours per day, 400 days per year, would have accumulated some 4×10^{12} elementary bits of information in a lifetime of 100 years.

mechanical design to be concerned more with mechanisms for the rapid repetition of ultrastereotyped tasks than with the development of highly flexible or universal mechanisms. The coordinating "mind" computers furnish thus finds no really suitable "body" ready to receive it. Nevertheless, the increasing use of computers in industrial and military process control and the association of computers with various mechanical and electric effectors in "real-time" scientific experimentation have begun to extend our experience with the problem of attaching muscles to the computer brain. A few interesting special research efforts in this area are also under way.

4. A number of interesting experiments have shown that computers can learn if they are programmed with criteria of success in performance and with ways to use these criteria to modify the parameters determining behavior choice in situations too complex for exhaustive calculations to be feasible. Such a technique has been used successfully to train character-reading programs to discriminate between letters printed in a given type face and to allow a checker-playing program to improve its ability by studying past games or by practicing against a copy of itself. The learned adaptations in these cases are, however, modifications of parameter in a program of fixed structure rather than acquisitions of more fundamental behavior patterns. An important basic method in natural learning is the discovery of significant similarities in situations that are not nearly identical. This fundamental activity of the mind has not been duplicated on the computer in any striking way. We do not yet have methods whereby the vitally important process of learning basic general principles from fragmentary indications can be made to function effectively.

5. A structure of motive and emotion, ultimately serving the purpose of individual and group preservation, necessarily forms a part of the living brain. Motives of preservation and predation have perhaps been the principal forces in the evolution of intelligence. (Plants, not being predators, have never developed intelligence.) Nevertheless, it is the ability to perform a set task successfully and rapidly, not the will either to perform it or not to perform it, which measures intelligence. At any rate, as computers grow in ability and steadily come to occupy more strategic positions in economic and social function, it becomes more desirable that they have no programmed motive other than that of advancing the general purposes of the institution they serve.

6. The elements of intelligence mentioned are all part of the inheritance we share with other mammals and even with animals generally. Certain additional aspects of intelligence, while existing in relatively primitive forms in animals other than man, are much more highly developed in man. These may perhaps be typified by the understanding of language and the ability to deal with abstract concepts.

A great deal of work has gone into developing the linguistic abilities of computers. Much is now understood about how underlying grammars permit the reconstruction of relationships implicit in a sequence of words or symbols. Such understanding is currently applied in a routine way to the interpretation of mechanical "source languages" with narrowly limited vocabularies and somewhat restricted grammars. Languages of this kind have become the standard vehicle of man-computer communication. Work on natural languages, with their large vocabularies and considerably more elaborate grammars, has also gone forward with fair success, constrained, however, by the complexity of the resulting programs and by the limitations of computer memory. Such work, incidentally, would benefit greatly from any substantial progress in the programming of computer learning, which would make it possible for computers to acquire their grammar, as people do, by studying source text. There is, all in all, no reason to anticipate fundamental difficulties in developing the computer's ability to understand language as far as resources allow.

Arithmetic and Boolean logic form the computer's basic reflexes, so that there is a real sense in which the computer's ability to deal with abstract concepts is inherent. Experiments in theorem proving by computer may be considered as explorations of this issue in a context of reasonable complexity. These studies have succeeded in devising methods whereby the computer can manipulate general concepts, discover the proofs of general statements, and even guide its conceptual explorations by heuristic principles of elegance and effectiveness. Nevertheless, they founder relatively quickly because of the computer's present inability to sense useful similarities in sufficient breadth. This, rather than any difficulties associated with abstractness *per se*, seems to be the principal block to improved performance. If a computer could find worms on a twig with the effectiveness of a bird, it might not fall so far short of the mathematician's ability to hunt out interesting theorems and definitions.

Prospects at Long Term

Humanity is constrained, not in its tools but in itself, by the infinitesimal pace of evolution. If we are able to create a human intelligence, we shall shortly thereafter be able to create highly superhuman intelligences. These may have the form of huge self-running factories producing varied objects in response to a stream of incoming orders; or of groups of roving sensor-effectors, gathering external visual or other data, reducing them for transmission and subsequent analysis to a central brain, and acting in accordance with retransmitted instructions, the whole forming a kind of colony or army of telepathic ants; or of an electronic mathematician, selecting interesting consequences from the association of imaginations at a superhuman rate.

It is apparent that such devices would have an enormously profound effect on all the circumstances of human society. It seems to the author that present computers are, in fact, the earliest species of a new life form, which may be called *crystallozoa* in distinction to the older *protozoa*. In the near future, on a historical scale, the crystallozoa will be able to store information with the same molecular-level density as do the protozoa. They will then enjoy advantages corresponding to the 10^6:1 ratio of electronic/protoplasmic clock rates or to the 10^4:1 ratio between the speed of light and the speed of signal transmission along neurons. If any of this is true, it is clear that what has been seen until now is but the faintest indication of what is to come.

BIBLIOGRAPHY

Marvin Minsky, *Computers: Finite and Infinite Machines*, Prentice-Hall, Englewood Cliffs, N.J., 1967.

Martin Davis, *Computability and Unsolvability*, McGraw-Hill, New York, 1958.

E. Feigenbaum and J. Feldman, *Computers and Thought*, McGraw-Hill, New York, 1963.

Dean E. Wooldridge, *The Machinery of the Brain*, McGraw-Hill, New York, 1963.

David H. Hubel and T. N. Wiesel, "Receptive Fields of Single Neurons in the Cat's Striate Cortex," *Journal of Physiology*, *148*, 574–591 (1959).

BIOGRAPHICAL NOTE

Jack Schwartz, author of two of the essays in this volume, is a professor at New York University's Courant Institute of Mathematical Sciences. He first became involved in functional analysis during his graduate studies at Yale University and since then has written a number of research papers in this field. Although he has been active in the computer field for only a few years, his interest in it actually dates back to his undergraduate days at the City College of New York, when he did some work on computer-related logic.